Abraham De Moivre

Abraham De Moivre

❧ Setting the Stage for Classical Probability ❧
and Its Applications

David R. Bellhouse

CRC Press
Taylor & Francis Group
Boca Raton London New York

CRC Press is an imprint of the
Taylor & Francis Group, an **informa** business
AN A K PETERS BOOK

CRC Press
Taylor & Francis Group
6000 Broken Sound Parkway NW, Suite 300
Boca Raton, FL 33487-2742

First issued in paperback 2019

© by Taylor & Francis Group, LLC
CRC Press is an imprint of Taylor & Francis Group, an Informa business

No claim to original U.S. Government works

ISBN-13: 978-1-56881-349-3 (hbk)
ISBN-13: 978-0-367-38225-4 (pbk)

Visit the Taylor & Francis Web site at
http://www.taylorandfrancis.com

and the CRC Press Web site at
http://www.crcpress.com

To Louise

Table of Contents

Preface

In a massive article on probability theory written for the seventh edition of the *Encyclopaedia Britannica*, the mathematician and actuary Thomas Galloway commented:

> About the same time [when Pierre Rémond de Montmort's and Jacob Bernoulli's works on probability appeared in 1708 and 1713, respectively], Demoivre began to turn his attention to the subject of probability, and his labours, which were continued during a long life, contributed greatly to the advancement of the general theory, as well as the extension of some of its most interesting applications.[1]

A few pages later Galloway adds:

> Since the time of Demoivre, the English treatises on the general theory of probability have neither been numerous, nor, with one or two exceptions, very important. Simpson's Laws of Chance (1740) contains a considerable number of examples, in the solution of which the author displays his usual acuteness and originality, but as they belong entirely to that class in which the chances are known a priori, they give no idea of the most interesting applications of the theory. Dodson's Mathematical Repository contains a large selection of the same kind. The Essay in the Library of Useful Knowledge, by Mr. Lubbock, gives a more comprehensive and philosophical, though an elementary view of the subject; but by far the most valuable work in the language is the Treatise in the Encyclopedia Metropolitana, by Professor De Morgan, 1837. In this very able production, Mr. De Morgan has treated the subject in its utmost generality, and embodied, within a moderate compass, the substance of the great work of Laplace.

One gets the distinct, and correct, impression from these quotations that Abraham De Moivre was Britain's leading probabilist of the eighteenth century. And not much else occurred in British probability until the nineteenth century. I will hold to that last sentence and later expand on it, despite possible protests from the fans of Thomas Bayes. Many, including me, have chased Bayes,[2] an obscure Presbyterian minister who had no impact on the subject of probability in his own lifetime and only a little afterward. Few have given Abraham De Moivre the treatment he truly deserves.

The first part of the quotation that I have given from Galloway underlines the secondary theme of this book. De Moivre began working as a pure mathematician. As he became established, he moved into applications, to finance in particular, where he made some fundamental contributions. His work in probability and its applications was often motivated by questions that his friends, patrons, colleagues, and clients posed to him. This is the major theme of the book. The eighteenth century operated on the connections, professional and personal, that a person established with others. De Moivre had a fairly wide network of friends and patrons and often benefited from it. In describing his life and work, I will elaborate on De Moivre's connections, which have usually stayed below the surface in many treatments of his work.

I was motivated to write this book after spending the past twenty-five years studying the development of probability in Britain. One of my favorite projects was collaborative work with Christian Genest.[3] We spent many pleasant hours examining and expanding on Matthew Maty's original biography of Abraham De Moivre.[4] Since that time I continued on my own and found even more material related to De Moivre's life and work.

Greatly encouraged by Stephen Stigler, I first tried to write a book on the development of probability in eighteenth-century Britain. After getting about halfway to completion, it became more than obvious to me that what I was writing was a prelude to De Moivre's work, a description of his work, and a postlude that looked at the impact of his work in Britain. It made more sense to devote the entire book to De Moivre. What I present here are the fruits of my labor over the past three years or so.

In this new effort I had a problem. In 1968, Ivo Schneider wrote a very lengthy article in German about De Moivre and his work.[5] I do not read German at all, so most of his paper was inaccessible to me. I was able to access some of the material through those who understand German, such as Anders Hald. They have interpreted Schneider's work in their English publications on the history of probability.[6] There was some advantage to my lack of German. I was forced to read and interpret what De Moivre had written, independently, for the most part, of someone else's interpretation.

De Moivre wrote a number of articles in Latin. My surviving Latin from high school is limited but much greater than my nonexistent German. I was greatly helped in this area in several ways. The first is Bruce McClintock's translation of *De Mensura Sortis*.[7] The second is that several of De Moivre's articles, or parts of them, were translated in the eighteenth century and appear in publications called *Miscellanea Curiosa*, written to give the general public of the eighteenth century access to papers in *Philosophical Transactions*. The third is that I have been blessed with a graduate student, Elizabeth Renouf, who decided after a degree or two in classics to follow a career in statistics. Finally, during an internet search I came across a French publication that I eagerly purchased. It is a translation into French by Jean Peyroux of De Moivre's *Miscellanea Analytica*[8] and it has been very useful to me. My French is much better than my Latin. And so with the French translation in one hand and the Latin original in the other, accompanied by a French-English dictionary on my desk in case of emergency, I was able to get a good grasp of De Moivre's work in his *Miscellanea Analytica*.

Although De Moivre left his papers to his friend George Lewis Scott, these papers apparently have not survived. Consequently, the information cobbled together on De Moivre's life comes from many sources, some published (including some of De Moivre's letters to others) and some from manuscripts held in various libraries in Britain and elsewhere. Collecting the information has been a paper chase that has been both challenging and rewarding. I have spent many pleasant hours in libraries and have corresponded with many helpful librarians. I have also benefited enormously from Google Books which has allowed me to read many obscure eighteenth-century publications while sitting in my office.

Acknowledgments

Putting together this biography has involved research work, either in person or through the librarian on the spot, in many libraries in England, France, Germany, Russia, Switzerland, and the United States. I am grateful to all those librarians who have given me so much help. Iris Lorenz at Staatsbibliothek zu Berlin was very helpful in providing me with a copy of a De Moivre manuscript. Fritz Nagel of Universitätsbibliothek Basel has made a large amount of the Bernoulli correspondence accessible via the Internet. There have been many others, and I regret to say that I did not record their names. For work in one library, La Bibliothèque de la Société de l'Histoire du Protestantisme français, I must single out Antoine de Falguerolles who went to the library for me to locate a genealogy of the Moivre family once I had found a reference to it. I would also like to thank Menso Folkerts who lent me his microfilm copy of a De Moivre manuscript that is in Landesbibliothek Oldenburg. Concerning a De Moivre manuscript that used to be held in a Russian library, I would like to thank Ksenia Bushmeneva who talked to several Russian and Polish librarians in an attempt to locate the manuscript (now destroyed).

Translation from French to English of some source material that appears in the article that I wrote with Christian Genest is due to Genest. In a few cases, I have given my own translation from French to English of new material that are extracts from letters. When the subtlety of the language was beyond me, which was in most cases, I relied on Catherine Cox. She is a very able translator who has done much work for the Statistical Society of Canada.

Several images appear in the book. I would like to thank the Royal Society, the Huguenot Society of Great Britain and Ireland, the British Museum, the Victoria and Albert Museum, the Getty Research Institute, the Canadian Centre for Architecture, the Buckinghamshire County Museum, and St. John's College, Cambridge for allowing images in their possession to be reproduced here.

It was Stephen Stigler who first gave me encouragement when I began to work in the history of probability and statistics many years ago. He has continued to provide his encouragement as well as highly insightful comments about things I have written. Reproductions of the allegorical engravings in Montmort's *Essay d'analyse* and De Moivre's *Doctrine of Chances* come from books in his collection. When I was working on the dispute between Simpson and De Moivre and found something odd about De Moivre's later editions of *Annuities on Lives*, Stigler graciously examined his own original copies to confirm my suspicions.

Along the way I have met several historians at conferences and elsewhere who keep reminding me to put my histories of technical subjects into their proper historical context. I hope that I have been successful in following their advice by writing this book. First among these historians is Bill Acres of Huron University College, whom I first met by sitting in on his class on Renaissance history. Since then we have collaborated on historical research projects that have enhanced my insight into the life of Abraham De Moivre.

In the course of writing this book I have relied on the proofreading skills of Elizabeth Renouf and Matthew VanderHeide. They have greatly improved the presentation of what I have written. With her background in classical studies and statistics, Elizabeth has also been very helpful with some Latin passages. Finally, I would like to thank Christian Genest and George and Evelyn Styan who, with their keen eyes and skills as editors, have provided comments on and corrections to the first draft of this biography.

✤ 1 ✤
Early Life in France

By the end of his life, Abraham De Moivre was viewed by some of his contemporaries as a religious skeptic.[1] It was, however, religion that shaped the course of his early life and was the source of disruption in his mathematics education. The Moivre[2] family were Huguenots, Calvinist Protestants in a predominantly Catholic France.

The Reformation of the sixteenth century and the introduction of John Calvin's teaching into France resulted in intermittent persecution of Protestants beginning in the 1530s. This escalated as powerful political forces came into play. The French Wars of Religion, running from the 1560s to near the end of the century, were a struggle for power between the Catholic House of Guise and the Protestant House of Bourbon. The wars ended when the Protestant Henry of Navarre became King Henry IV of France and converted to Catholicism. In 1598 Henry proclaimed the Edict of Nantes, providing some religious freedoms to the Huguenots. The edict was revoked in 1685 by his grandson, Louis XIV, as he centralized the power of the state and control over the Roman Catholic Church on himself. Louis made it illegal to practice as a Protestant, and all Protestant churches and schools were closed.

Abraham De Moivre was born on May 26, 1667, in Vitry-le-François,[3] a town situated about 175 kilometers east of Paris in the province of Champagne (as it was known under the Ancien Régime).[4] In 1544 the ancient town of Vitry had been put to the torch during one of the military clashes between the Holy Roman Emperor Charles V and François I, the reigning French king. The next year François ordered a new town to be built nearby according to the latest ideas in Renaissance town planning and fortification.[5] The new walled town, laid out on a rectangular grid system, was named Vitry-le-François after the king. From a military standpoint, it was strategically placed to command two major transportation routes, although the

Vitry-le-François in 1634 (from the Canadian Centre for Architecture collection).[6]

fortifications were not completed until the 1580s. Since Vitry-le-François sat at the crossroads between Paris and major centers to the east, including Switzerland, it was soon exposed to the new religious ideas coming out of Switzerland. By the early 1560s there was a Huguenot pastor resident in the town tending to a growing congregation.[7] At the time of De Moivre's birth, the town had a thriving and sizable Huguenot community. The population had also expanded beyond the walls of the original town, numbering about 12,000 in 1685.[8]

Prior to Abraham's birth, the Moivre family had lived in Vitry-le-François for a few generations. Abraham's great-grandfather, Aggée de Moivre, worked in the leather industry. Leather tanning was a major industry in Vitry-le-François; there were at least ten tanneries in the town prior to 1685.[9] With continued success of the tanning industry in Vitry-le-François, the following generation of Aggée de Moivre's family became solidly part of the merchant class. Abraham's grandfather, Daniel de Moivre, was a merchant; Abraham's father, also Daniel, was a surgeon. Another relative, Jean de Moivre was a portrait painter, executing many portraits of Protestants in Vitry-le-François in the early 1640s.[10] On May 25, 1665, when he was 37 years of age, Abraham's father married Anne Bureau of Paris at the Protestant church in Heiltz-le-Maurupt,[11] a town about 22 kilometers from Vitry-le-François. About two years after Abraham's birth in 1667, his brother Daniel was born.

As youngsters, the Moivre children were originally educated in Vitry-le-François. When they first entered school, Protestant education in France was already in a state of flux.[12] As it developed in the sixteenth century, the Huguenots had a school system that extended from primary education to the university level, the petite école to the académie. The primary curriculum included both religious instruction and education in Latin and Greek authors, culminating in the study of rhetoric through classical works such as Cicero's *De Oratore*. From there the student could progress to higher studies in the academies by taking logic, ethics, and metaphysics. Students destined for the ministry received theological training in the academies. Government-imposed restrictions on the Protestant education system began in 1670, when Abraham De Moivre was three years old. In that year, teachers in the Huguenot primary schools "were forbidden to teach any subjects beyond reading, writing, and arithmetic."[13] Only one teacher was allowed per school and there were restrictions on enrollment that made it difficult for students to study in a Huguenot school if it did not already exist in their hometown. Restrictions and regulation continued to tighten after 1670 until 1685, the year of the revocation of the Edict of Nantes. In 1685 any remaining Huguenot academies that had survived this restrictive atmosphere were closed.

For twenty years or more prior to Abraham De Moivre reaching school age, the town school in Vitry-le-François underwent several administrative reorganizations. Known as the Collège de Vitry, both Protestant and Catholic children were taught at the school. For over 45 years it was run by a Protestant layman, Jean Garnier. In 1649, when Garnier became too old to handle his duties, the city decided to contract out the teaching. Initially, it was to the Catholic Order of Minims. After some scandals over student behavior, the contract next went to the Oratorians, another Catholic order. In 1665, there was a Catholic mission to Vitry-le-François to convert Protestants. The Collège de Vitry was turned over to the Fathers of the Christian Doctrine (*Pères de la doctrine Chrétienne*).[14] This was a Catholic teaching order of priests, originally founded in 1592 during the Counter-Reformation to instruct students in Roman doctrine and to stem the spread of Calvinism,[15] the very theology underlying French Protestantism. The Fathers of the Christian Doctrine tried to convert their Protestant students, an act which made many Huguenot townspeople unhappy. Very soon the Fathers came to an agreement with the citizens that Protestant students would be taught the same as other children in the school, but would be excused from all prayers, instruction in the Catholic catechism, and other religious exercises.[16] Abraham De Moivre initially took lessons in Latin from a local Catholic priest.[17] After a year with the priest, he moved on to receive instruction from the Fathers of the Christian Doctrine at the Collège de Vitry. Apart from this schooling, De Moivre also received private tutoring in arithmetic. This situation continued until about 1678 when De Moivre's father, dissatisfied with the arithmetic tutor as well as with the school run by the Fathers of the Christian Doctrine, sent his son to the Protestant Academy in Sedan, a town near the Belgian border northeast of Paris and about 115 kilometers north of Vitry-le-François. There Abraham transferred into the traditional education system of the Huguenots.

Although many laypeople like De Moivre attended the academy, its raison d'être was the training of theology students for the ministry. Of the ten faculty members, three were in theology. The faculty complement at the academy in Sedan was about ten. They taught subjects in theology, Hebrew, Greek, jurisprudence, mathematics, philosophy, and rhetoric. The study of Hebrew and Greek was related to theological training. Students were expected to be able to read the Old and New Testaments in their original languages. The day-to-day life of the students also had a religious foundation. It was the norm for students to board with the professors or with the regents of the academy. This would ensure that students said their prayers every morning and evening in the home. Classes, which were opened with prayer, went from Monday to Friday, about seven hours a day, with only a few weeks of vacation in the year. Wednesdays were devoted mainly to sermon writing, presumably by the theology students. On Saturdays students typically spent their time studying the catechism, making disputations in Greek or Latin, or reciting what they had learned during the week.[18]

The more eminent professors at Sedan in the late 1670s were Pierre Bayle, who taught philosophy and history, Jacques du Rondel, who taught rhetoric and Greek, and Pierre Jurieu, who taught theology.[19] The faculty complement may have been reduced during De Moivre's time due to the restrictions and pressures imposed by the Royal authorities. For example, there is no record of a mathematics professor at Sedan after 1672.

The system divided studies into six grades, one being the most senior and six the most junior. These were not grades by year, but grades by mastering certain topics. By the time De Moivre reached Sedan he had already completed the third grade of his studies. In terms of the curriculum at Sedan, he would have already mastered topics in Latin grammar, some Latin authors such as Ovid and Cicero, and an introduction to Greek. The second grade at Sedan consisted of Greek for New Testament studies, more Ovid as well as Virgil, and rhetoric.

At Sedan, De Moivre studied Greek in his first year and followed that with rhetoric in the next under Rondel. He may also have had some lessons in arithmetic from Rondel. Matthew Maty, Moivre's first biographer, provides an anecdote told to him by De Moivre, showing the mathematician's arithmetical interests at Sedan.

> Whenever his teacher [probably Rondel], who was not so keen on arithmetic as he was on Greek, found the table of his pupil forever strewn with calculations, he could not help wondering *what does this little rogue intend to do with those numbers?*[20]

De Moivre was about thirteen years old at the time.

Despite the lack of a regular mathematics professor at Sedan, De Moivre studied François le Gendre's book *L'Arithmétique en sa perfection* with the help of a fellow student. First published in 1648,[21] *L'Arithmétique* was probably the best known French commercial arithmetic book. It went through many editions and continued to be in print into the early nineteenth century. The book contains some of the elementary topics that De Moivre taught as a tutor in mathematics after his arrival in England. There are descriptions of the common practical methods of addition, subtraction, multiplication, and division by hand, including methods for whole numbers and fractions. Also covered are the extraction of square roots and cube roots, as well as methods such as the rule of false position and the rule of three. After covering the basics, there are discussions of practical business arithmetic, again with worked-out examples. Topics in this section include finding the value of one country's or area's currency in terms of another's (*francs bordelois* into *livres de gros* of Flanders and vice versa, for example) and translating the weights and measures of one country or area into those of another. The topics related to geometry are devoted to finding the areas of plane figures: triangles, rectangles, trapeziums, and rhombuses. What may have caught De Moivre's attention is a section containing challenge problems. These questions appear to build on the reader's knowledge with examples that are different from the usual problems in the book.

In 1680, De Moivre returned to his home in Champagne. He intended to return to Sedan to study with Pierre Bayle, but in 1681 the academy was suppressed by the French authorities, who turned the building over to the Jesuits for a fee of 20,000 livres. Many of the faculty members left France, as was to become typical after the closure of the Protestant academies. Bayle and Jurieu took refuge in Rotterdam where they obtained teaching positions at the École illustre; Rondel went to the University of Maastricht. Soon Bayle and Jurieu had a serious falling out. (Jurieu accused Bayle of treasonous activities.)[22] Bayle was removed from his position but remained in Rotterdam for the rest of his life.

At home, De Moivre tried self-study in mathematics by reading Jean Prestet's *Éléments de mathématiques*, but, at the age of only fourteen, he had difficulty understanding some of the initial concepts in the book. As Maty relates,

> Unfortunately, the young man found in the introduction to this treatise a preliminary discussion on the nature of our ideas, and since he did not know what an idea was—he had never had the good fortune to hear Mr. Bayle on the subject—he closed the book without ever reading it.[23]

With Sedan closed, De Moivre then went to the Protestant Academy at Saumur in 1682. It was situated about 385 kilometers southwest of Vitry-le-François. Like Sedan, Saumur had upwards of nine professors covering several areas of the humanities. Like Sedan, there is no surviving record of a mathematics professor at Saumur during De Moivre's time there.

From 1682 to 1684 while he was at Saumur, De Moivre studied logic and mathematics, finally coming to grips with Prestet's work. Despite the absence of a mathematics teacher, it was at Saumur that De Moivre came across a copy of Christiaan Huygens's 1657 treatise, *De Ratiociniis in ludo aleae*, and read it.[24] This work, the first published work in probability theory, came about as a result of a visit that Huygens made to Paris in 1655. There Huygens learned of the correspondence between Blaise Pascal and Pierre de Fermat regarding the classical probability problems of the day: the division of stakes and problems related to the throw of dice. Probably motivated by the challenge of the difficulty of these problems, Huygens solved them himself and then proceeded to write his treatise, originally in Dutch and then translated into Latin by his teacher Frans van Schooten. In addition to providing solutions to the two classical probability problems, Huygens set out five challenge problems at the end of the book accompanied by numerical answers.[25]

De Moivre met two teachers at Saumur who almost certainly might have been of assistance to him in England once they had all fled France. These were Abraham Meure, who taught Greek and rhetoric, and Jacques Cappel, who taught Hebrew.

At Saumur, De Moivre was introduced to ideas in physics put forward by René Descartes. De Moivre's teacher in physics at Saumur, known only as a Scotsman named Mr. Duncan,[26] was not satisfactory as a teacher. And so in 1684 De Moivre went to Paris to study physics at the Collège d'Harcourt, a college at that time associated with l'Université de Paris. The textbook that De Moivre most likely would have used in his studies at Saumur and Harcourt was Jacques Rohault's *Traité de physique*, first published in 1671. It went through several editions, and soon after its initial publication became the standard physics textbook in both France and England. Rohault was an expositor and systematizer of Descartes' approach to physics.[27] Compared to Isaac Newton's *Principia Mathematica*, which was published in 1687, Rohault's approach to physics was very nonmathematical.

It was Descartes' approach to celestial mechanics with which Newton clashed in his *Principia*. Some key elements of the Cartesian system were that matter is indivisible and impenetrable. Consequently, the existence of a vacuum was not possible and so planets travelled in their orbits not because of gravitational attraction, but through vortices. French scientists in the late seventeenth and early eighteenth centuries spilled much ink trying to justify the necessity of this swirling cosmic matter that would allow the planets to move in their orbits.[28]

After a year at Harcourt, De Moivre once again returned home and then went to visit relatives in Burgundy. He continued his mathematical studies on his own by reading the first six books of Euclid and other works in elementary plane geometry. During his initial reading of Euclid, he got stuck on Proposition V in Book I, which states that the angles at the base of an isosceles triangle are equal. It is the first difficult proposition in Euclid and was known as the *pons asinorum* (bridge of fools) since that is where many initiates to Euclid come to grief with their studies of geometry. One of De Moivre's Burgundian relatives helped him with the proposition and the rest of Euclid became smooth sailing. Upon finishing Euclid, De Moivre went on to topics in trigonometry, perspective, mechanics, and spherical triangles.

It was at this time, as part of the increasing pressure on Huguenots exerted by the authorities, that Huguenot surgeons such as De Moivre's father were prohibited from practicing their profession.[29] Shortly thereafter, Louis XIV revoked the Edict of Nantes, thus suppressing Protestant education throughout France as well as closing

Protestant churches. The Protestant church at Vitry-le-François, for example, which had been built in 1613, was acquired by the Fathers of the Christian Doctrine in 1685. The other four Protestant churches in the area around Vitry-le-François were torn down.[30] Protestant ministers who did not convert to Catholicism were expelled from the country, while the laity was forbidden to leave on pain of being sent to the galleys (essentially prison ships). Despite the restrictive law on emigration, thousands fled to other countries.

Accompanied by his father, Abraham De Moivre returned to Paris where he was tutored in mathematics by Jacques Ozanam. At the time Ozanam was about 45 years old and was making a comfortable living at tutoring and writing mathematics books, mostly for instructional purposes. In terms of social class, De Moivre, coming from the merchant class, was a typical Ozanam student.[31] In terms of mathematical ability, De Moivre claimed, at least late in life, that he was superior to Ozanam in mathematics.[32] Under Ozanam, De Moivre studied the geometry of solids: cones, pyramids, cylinders, and spheres. He also honed his chess-playing skills with Ozanam, a skill that, once he arrived in England, was useful to him when interacting with some of his aristocratic chess-playing patrons.

In Vitry-le-François the Moivres had connections to some local government officials who were fellow Protestants. These connections become apparent after the revocation of the Edict of Nantes. There was an individual in Vitry-le-François named Claude de Marolles who was interested in mathematics, specifically algebra. His kinsman, Louis de Marolles was an official in the courts of justice in Sainte-Menehould, about 55 kilometers from Vitry-le-François.[33] The extended Marolles family was one of the three leading Protestant families in Champagne;[34] Claude was one of the leaders in the Protestant church in Vitry-le-François, which the Moivre family would have attended. De Moivre was connected to the Marolles through mathematics. Claude taught Louis algebra at which Louis became very adept, to the point that he wrote a manuscript treatise on algebra containing many difficult problems. After the revocation of the Edict of Nantes, Louis de Marolles was one of those caught and arrested while trying to leave France. Prior to sending him to the galleys and certain death, an attempt was made to convert Marolles to Catholicism; presumably the conversion of a senior member of a prominent Protestant family would make good propaganda for the state. In June or July of 1686 Marolles was sent to La Tournelle in Paris, a former royal palace turned into a prison. The young De Moivre visited Marolles at La Tournelle, probably as a friend of the family bringing comfort and help to the prisoner. During Marolles' confinement, he so consistently refused to convert that the clergy in charge of him began to spread rumors that he had lost his mind. To show that he was not insane, Marolles proposed a mathematical problem that he would solve. De Moivre later reported that the problem was taken from one of Ozanam's books.[35]

De Moivre's father, Daniel, was also detained in Paris, but somehow was able to send his family to England. His final imprisonment occurred on December 9, 1687, on the orders of the Marquis de Seignelay, one of the ministers in Louis' government and son of the famous Jean-Baptiste Colbert. Daniel de Moivre (described only as "le chirurgien Moivre") was held at the Prieuré de Saint-Martin-les-Champs, a location in Paris different from where the authorities held Marolles.[36] He was released April 27, 1688, which meant that the authorities had broken him and he had recanted his Protestant beliefs. Daniel de Moivre died in Paris; he would have been about sixty years of age. This situation lends credence to an anecdote attributed to Abraham De Moivre and recounted about one hundred years after it may have occurred. The story runs,

> I have heard my father say that De Moivre being one day in a Coffee-house in St. Martin's Lane, much frequented by Refugees and other French, overheard a Frenchman say that every good subject ought to be the religion of his King—'Eh quoi donc, Monsieur, si son roi professe la religion du diable, doit-il suivre?' [Well then, Sir, if his king professes the religion of the devil, should he follow him?][37]

(There are some inaccuracies in the full anecdote, so its reliability is in question.[38])

During the incarceration of his father, Abraham De Moivre, along with his brother Daniel and his mother Anne, fled France across the English Channel despite Louis' laws prohibiting emigration of Huguenots.[39] After their arrival in England, the two young men, aged 20 and 18, presented themselves to the Savoy Church, one of the two main French Protestant churches in London. It was situated in The Strand, near where Abraham De Moivre eventually lived. On August 28, 1687, Abraham and Daniel made their reconnaissance at the Savoy Church, or reaffirmed their loyalty to the Protestant faith.[40] In France they had been forced to renounce their Protestant faith and attend Roman Catholic Church services.

The young refugees never saw their father again. The father, in detention when the rest of the family fled, was only released well after mother and sons arrived in England. He died in Paris shortly after his release from the Prieuré de Saint-Martin-les-Champs before he could be reunited with his family. His wife, Anne, lived for another twenty years in London with the younger son, Daniel.

When the two brothers made their reconnaissance at the Savoy Church, they signed a document saying that they would be faithful to their reaffirmed Protestant beliefs. Though their signatures are side-by-side as shown here from the book of reconnaissance, they are different, showing two distinct variations. Abraham used an uppercase letter to begin the particle "de" in his name. Daniel used the usual lowercase in the particle as well as in his given name. Also, Daniel used a typical variant spelling of the surname using a *y* instead of *i*. Since the older brother most often spelled his surname "De Moivre," I will adhere to this throughout the book, rather than the strictly correct French usage "de Moivre" or just plain "Moivre."

From an early age, Abraham De Moivre showed a great aptitude in mathematics, as shown by anecdotes of his experience at the academies in Sedan and Saumur,

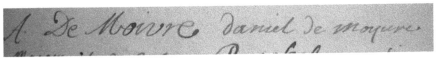

Earliest known signatures of Abraham and Daniel De Moivre.[42]

followed by his course of study with Jacques Ozanan. His education in mathematics was a mixture of self-taught topics and topics learned from tutors. Despite the interruptions due to Louis XIV's persecution of Protestants, his education prepared him well for his future life in England. There he established himself as a tutor in mathematics and was very quickly admitted to England's scientific society.

❧ 2 ❧

Points of Connection

There are two types of connections that are relevant to Abraham De Moivre's career: personal and intellectual. Personal connections are defined by the network of friends, acquaintances and colleagues that are built socially, professionally, and through the workplace. Certainly Abraham De Moivre met and interacted with numerous people during his lifetime, many of whom had a significant impact on his career. I would suspect that when asked to name De Moivre's associates, historians of science would probably place Isaac Newton at the top of their list. There are many others on that list including Huguenot friends, fellows of the Royal Society, and members of landed families who acted as patrons for his teaching career and for his scientific publications. Intellectual connections are the chains of ideas that come together to produce new knowledge. Again, Newton, or more correctly, Newton's work in mathematics and in physics (or what was called natural philosophy in Newton's day), figure prominently among the ideas that stimulated De Moivre's mathematical work. What can stimulate new knowledge is the bringing together of seemingly disparate ideas. Where Abraham De Moivre was most successful was in developing new areas of probability theory. There he melded together some traditional approaches to the calculation of probabilities with other areas of mathematics that typically had not been used in probability theory before. Newton's binomial theorem and his work in infinite series were major stimuli for De Moivre's work in probability.

Personal connections and intellectual connections are not separate entities that operate independently. Ideas are usually discussed between individuals—in direct conversation, in letters, and in print. The communication network for intellectuals that was operational during De Moivre's lifetime is known as the Republic of Letters. A succinct description of this network appears on a website for a project that is trying to map the network connections in the Republic of Letters from about 1500 to 1800.

> The Republic of Letters was an intellectual network initially based on the writing and exchange of letters that emerged with and thrived on new technologies such as the printing press and organized itself around cultural institutions (e.g. museums, libraries, academies) and research projects that collected, sorted, and dispersed knowledge. A pre-disciplinary community in which most of the modern disciplines developed, it was the ancestor to a wide range of intellectual societies from the seventeenth-century salons and eighteenth-century coffeehouses to the scientific academy or learned society and the modern research university. Forged in the humanist culture of learning that promoted the ancient ideal of the republic as the place for free and continuous exchange of knowledge, the Republic of Letters was simultaneously an imagined community (a scholar's utopia where differences, in theory, would not matter), an information network, and a dynamic platform from which a wide variety of intellectual projects— many of them with important ramifications for society, politics, and religion— were proposed, vetted, and executed.[1]

As his scientific status grew, De Moivre entered into the Republic of Letters. Unlike some of the mathematical giants of his day, such as Johann Bernoulli and Gottfried Leibniz who left large collections of correspondence, De Moivre appears to have been only moderately active in the Republic of Letters. This may be partly attributed to the reality De Moivre faced in making a living through private tutoring.

Personal Connections

Eighteenth-century English society was all about making personal connections and keeping them. To survive and get ahead, individuals needed to establish contacts with other people who could help them in some way. As one historian of the eighteenth century has put it, people fit into the English social strata of the time

> by their personal connections with others, especially authority figures: fathers, masters, husbands, parsons, patrons.... People had to shift for themselves. There was no all-encompassing welfare estate, no comprehensive system of social services, guaranteeing care from cradle to grave. How one made out depended on skills in the games of deference and condescension, patronage and favour, protection and obedience, seizing opportunities and making the most of them.[2]

Arriving with no English connections, De Moivre successfully survived and flourished within this system, first as a tutor to landed families and then as a consultant on annuity valuations.

When the Moivre brothers and their mother first arrived in London, the only people they likely would have known were fellow Huguenots who had fled France at about the same time or earlier. Among this group would be relatives, former teachers from Sedan and Saumur, and fellow townspeople from Vitry-le-François. One probable relative is Judic Morel, who was likely from Vitry-le-François.[3] After her arrival in London, in 1693 she married Hugues Le Sage from Alençon. Hugues and Judic's son, John (or Jean) Le Sage, whom Abraham De Moivre once referred to as his cousin, became a prominent London silversmith.[4] In 1718 Le Sage was living in the upper end of St. Martin's Lane, which is near where the Moivre brothers lived.[5] Another refugee from De Moivre's hometown is Isaac Garnier. He fled to England in early 1682, five years before the Moivre brothers. Originally trained in France in medicine and pharmacy, it is likely that Garnier knew Daniel De Moivre senior, a surgeon. Soon after Garnier's arrival in England, he was appointed to the position of Royal Apothecary under Charles II.[6] Two former teachers from Saumur, Abraham Meure and Jacques Cappel, made their living in London as teachers. Meure arrived in England with his family in 1685 or the next year.[7] Subsequently, he ran a school in Hog Lane (now the upper part of Charing Cross Road).[8] Cappel initially followed the more typical activity of someone in his station by becoming a tutor in an English household. Beginning in 1699 he tutored Martin Folkes for about seven years.[9] Folkes became a friend and student of Abraham De Moivre and later in life served as President of the Royal Society. After his time with the Folkes family, Cappel worked as a teacher at a dissenting academy in London.[10]

Throughout his life Abraham De Moivre made connections with several others in the Huguenot community. A list of some of his known Huguenot friends shows that they came from a wide variety of backgrounds, ranging from business to literary to military to political, with perhaps more emphasis on the literary. Pierre Des Maizeaux and Michel de la Roche were active in the Republic of Letters, where their paths sometimes intersected with De Moivre's. They were literary journalists who wrote reviews in French-language journals published in the Low Countries. These journals informed continental Europeans about newly published English books. They also wrote for other journals that made English audiences aware of French-language work. Pierre-Antoine Motteux, author and playwright, wrote for English audiences; his most noted works are translations of Rabelais and Cervantes. Another writer, Pierre Coste, did translation work for Isaac Newton and the philosopher John Locke. Matthew Maty, who wrote De Moivre's biography a year after the mathematician's death, was a writer and physician. Two years after De Moivre's death, Maty became a librarian in the British Museum. He founded *Journal Britannique*, one of the later French-language publications out of the Low Counties that reviewed English publications for a French audience. It was in *Journal Britannique* that Maty initially published his biography of De Moivre in 1755. Among other Huguenots of De Moivre's circle, Peter de Magneville and Isaac Guion were businessmen. The nature of Magneville's business is unknown, other than that he travelled extensively; Guion was a distiller. Peter Davall was a barrister and John Seguin was a cavalry officer in the Horse Guards.[11] Some of these individuals studied mathematics with De Moivre. Many were part of an émigré circle that met socially for intellectual discussions.

Other Huguenots that De Moivre knew were from the second generation. They had been able to take advantage of the patronage system that was part of the fabric of English society. Hector Berenger de Beaufain became Collector of Customs in South Carolina and Francis Fauquière, a director of the South Sea Company, became Lieutenant Governor of Virginia. Both were fellows of the Royal Society. Fauquière's father had worked for Newton at the Mint. The patronage system only went so far in rewarding the children of immigrants; both Berenger de Beaufain and Fauquière were sent to the colonies. Another Huguenot friend of De Moivre, Isaac Leheup, made enough money to be able to run for a seat and to sit as a Member of Parliament, but climbed no further up the patronage ladder.

There were two areas of London to which Huguenot refugees generally gravitated. One was the Leicester Fields/Soho area in the City of Westminster, which included three parishes: St. Anne Soho, St. Giles-in-the-Fields, and St. Martin-in-the-Fields. The Church of St. Martin-in-the-Fields is situated at the bottom of St. Martin's Lane. Today this is in the heart of London's West End theatre district. A short walk away, St. Anne and St. Giles are immediately to the northwest and northeast of St. Martin-in-the-Fields, respectively. In the eighteenth century, the area was part of the western suburbs of London. The other settlement area for Huguenots in London was Spitalfields in the eastern suburbs of London. The Spitalfields area was largely populated by textile workers, while in Westminster the Huguenots were associated with the more well-to-do. In the western suburbs there were Huguenots who manufactured luxury, items running the gamut from clocks to wigs and from guns to goldwork.[12] Some patrons of the Huguenots lived in the same area or close by. George Parker, 2nd Earl of Macclesfield, lived in the parish of St. Martin-in-the-Fields, while Charles Spencer, 3rd Earl of Sunderland, lived a short distance away in Piccadilly.

The Moivre brothers lived in these western suburbs among fellow Huguenots that included the Guion, Le Sage, and Meure families. At his death in 1754, Abraham De Moivre was living in the Parish of St. Anne.[13] This was the same general area where his brother Daniel De Moivre resided at his death in 1733.[14] Abraham was buried in the graveyard attached to the Church of St. Martin-in-the-Fields.[15] In 1707, Daniel was living in Earle (now Earlham) Street close to the Seven Dials, a road junction where seven streets converge. Located in the Parish of St. Giles-in-the-Fields, the Seven Dials is a short walk up St. Martin's Lane through St. Andrew's Street (now Upper St. Martin's Lane). When the astronomer Jerome Lalande visited England in 1763, he wrote in his diary, "J'ai vû le café de Slaughter où Newton allait tous les jours dans St. Martins Street, et la place de M. Moivre [I saw Slaughter's Coffehouse where Newton went every day in St. Martin's Street, and Mr. Moivre's rooms]."[16] From 1711 until his death in 1727, Newton lived in St. Martin's Street,

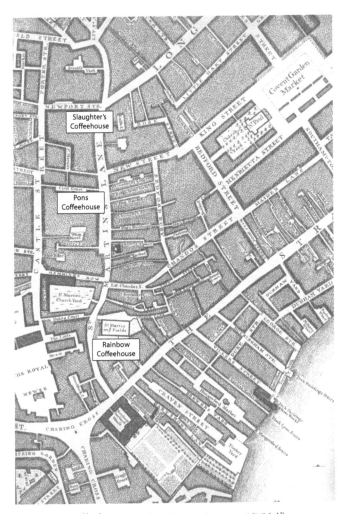

Coffeehouses in St. Martin's Lane in 1756.[17]

which runs into Leicester Square; Slaughter's Coffeehouse was nearby in St. Martin's Lane. A reasonable inference from Lalande's diary is that at the end of his life Abraham De Moivre lived in St. Martin's Lane, probably at the upper end since that area is part of the Parish of St. Anne. Unfortunately, there is no way to know for sure. Neither Abraham nor Daniel paid enough rent to show up in the Poor Law Rates, the taxation method of the time.[18] Abraham's lodgings were large enough, however, that he employed a servant by the name of Susanna Spella.[19]

De Moivre and his friends, whether Huguenot or not, met socially and for business purposes in coffeehouses. Abraham De Moivre himself frequented Slaughter's

Coffeehouse. He could also be seen at Pons Coffeehouse and the Rainbow Coffeehouse. Pons was situated in Cecil Court and the Rainbow in Lancaster Court, both off St. Martin's Lane. Their approximate locations are shown on the 1756 map of the St. Martin's Lane area.[20] Newton's house is off the map but is situated nearby, to the west of Pons Coffeehouse. Daniel's 1707 residence at the Seven Dials was next to Digory's Coffeehouse. The location of this coffeehouse is again off the map, but just above the end of St. Martin's Lane.

London coffeehouses of the eighteenth century might be viewed as "information central." There one could read one or more of the many London newspapers, exchange news verbally with other patrons, pick up one's mail, and obtain news and notices of new books and other publications.[21] Abraham De Moivre had his mail delivered to Slaughter's Coffeehouse. Sometimes called "penny universities," coffeehouses also served as informal centers of learning and debate. There is a story, probably apocryphal, that early in his career De Moivre gave lectures in coffeehouses on natural philosophy. He was unsuccessful in this endeavor, partly because of his poor English.[22] Coffeehouses were also places where a wide variety of business could be transacted. For example, subscriptions to a life insurance scheme could be obtained at Digory's Coffeehouse in 1710.[23] The same coffeehouse served as a consulting office in 1702 for a surgeon treating venereal disease.[24] Coffeehouses even served as lost and found depots.[25] Some coffeehouses specialized in and attracted certain clientele, perhaps the most famous being Lloyd's Coffeehouse, which evolved into a center for marine insurance. Pons Coffeehouse attracted many Huguenots, especially from the military, and the Rainbow Coffeehouse was where the Huguenot intelligentsia met, led by Pierre Des Maizeaux.[26] Patrons of Slaughter's Coffeehouse met to play chess and whist. When they were first published, copies of Edmond Hoyle's *A Short Treatise on the Game of Whist*, Philip Stamma's *The Noble Game of Chess*, and François-André Philodor's *Analyse du jeu des Échecs* were sold from Slaughter's.[27] These were the leading manuals on card games and chess in their time and for many years to come. Slaughter's was another haunt for Huguenots and, prior to the establishment of the Royal Academy of Arts in 1768, it was also a meeting place for artists.[28]

Coffeehouses were places where merchants, teachers and professional men could mix with the landed elites. In addition to Newton, Abraham De Moivre may have met Charles Spencer, 3rd Earl of Sunderland, at Slaughter's Coffeehouse. Sunderland was an avid chess player, one of the best in England.[29] De Moivre also played chess; he had honed his skills with Ozanam in Paris. He also applied his mathematical skills to chess by providing an elegant solution to the knight's tour problem—how to cover all 64 squares of a chess board using a knight's move without hitting the same square twice.[30] He probably solved this mathematical problem around 1718. The French mathematician and aristocrat Pierre Rémond de Montmort saw De Moivre's solution that year and found one of his own.[31] Although Sunderland was a fellow of the Royal Society, as was Newton and De Moivre,

Slaughter's Coffeehouses in the nineteenth century.[32] (© Trustees of the British Museum.)

Sunderland was an inactive member and may not have attended any meetings.[33] Instead, chess may have been De Moivre's entry into Sunderland's society.

During his career, De Moivre crossed paths with a number of aristocrats and landed families. He tutored many of this group's children in mathematics. The subscription list to his 1730 book, *Miscellanea Analytica*, is a who's who of early eighteenth century Whig politics in England.[34] Most notable among the names in the list is Robert Walpole, England's first prime minister and the ultimate player in patronage politics. Also included in the list is Walpole's son Edward as well as

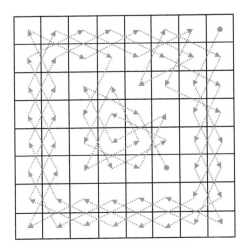

De Moivre's solution to the knight's tour problem.

Robert Walpole's close associate Charles, 2nd Viscount Townshend, and two of his sons, Thomas and William. De Moivre had connections that went into the upper reaches of British politics. His connections were also long lasting. Robert Spencer, 4th Earl of Sunderland, and Diana Spencer, later Duchess of Bedford, and both children of the 3rd Earl of Sunderland, also appear on the subscription list, although it is unlikely that either of them received tutoring from De Moivre. The 3rd Earl of Sunderland died at least six years before his children subscribed to *Miscellanea Analytica*; the children appear on the subscription list for no discernable reason other than their father's connection to De Moivre.

De Moivre was elected a fellow of the Royal Society in 1697. This provided him with another network of associates that included mathematicians and scientists, as well as aristocrats interested in science. De Moivre knew all the British mathematicians of his day, but was especially close to Edmond Halley, William Jones, Isaac Newton, and Brook Taylor. Compared to other mathematicians in the Royal Society at the beginning of the eighteenth century, Astronomer Royal John Flamsteed described De Moivre as "an honester and abler man than any of them."[35] Once plugged into the Royal Society network, De Moivre made connections within the international scientific community through the Republic of Letters. Prominent among De Moivre's international contacts were the Swiss mathematician Johann Bernoulli and his nephew Nicolaus Bernoulli, as well as the French mathematician Pierre Varignon.

Intellectual Connections

What occupied the minds of several mathematicians in the latter half of the seventeenth century and well into the eighteenth century were problems related to curves—finding the tangent to a curve at a given point, finding areas under curves (called quadrature), and finding lengths of curves (called rectification). The problems were all related through the newly developing differential and integral calculus that was discovered by Newton and his German counterpart, Gottfried Leibniz. On the Newtonian side are fluxions. Fluxions give tangents to curves. In the Leibnizian version of calculus, derivatives provide tangents. Newton's approach to fluxions is different from differential calculus today, which is more in tune with Leibniz's approach. Quadrature is to integral calculus as fluxions are to differential calculus. Once De Moivre arrived in England, one of his first mathematical interests was in finding new ways to obtain quadratures.

How Newton viewed curves, for example those of the form $y = (1 - x^2)^n$ that he studied in the mid-1660s, was in a very Euclidean way. Definition 2 of Book I of Euclid's *Elements* states, "A *line* is breadthless length."[36] By the late seventeenth century there were several commentaries on this definition. The first English translation of Euclid, done in 1570 by Henry Billingsley, provides a second definition in order to enhance the understanding of Euclid's original: "A line is the moving of a pointe, as the motion or draught of a pin or pen to your sense maketh a line."[37] Another commentary from 1685 provides more insight into motion and lines.

> 'Tis commonly said, that a Line is produc'd by the motion of a Point; which ought to be carefully observ'd; for motion may on that manner produce any quantity whatsoever; but here, we must imagine a Point to be only so mov'd, as to leave one trace in the space, through which it passes, and then, that trace will be a line.[38]

In Newton's mathematical worldview, a curve would be traced by the point of intersection of a horizontal and a vertical line moving through time.[39] From a modern viewpoint, the curve at time t during the tracing of it would be situated at the point $(x(t), y(t))$, so that Newton's variables x and y can be viewed as functions of time. In Newton's jargon, x and y are called fluents or flowing quantities; in this case they flow with time. The value of $x(t)$ is the distance at time t on the horizontal line from 0 to the point of intersection with the vertical line and $y(t)$ is the distance, also at time t, on the vertical line from 0 to the point of intersection with the horizontal line. The fluxion is the rate at which these fluents change over time. Newton denoted the fluxion of x as $\dot{x}$ and the fluxion of y as $\dot{y}$. In modern calculus notation, $\dot{x} = dx(t)/dt$ and $\dot{y} = dy(t)/dt$. The tangent to the curve is $\dot{y}/\dot{x}$, which in modern calculus notation would be the derivative dy/dx. The basic problems in Newtonian calculus are to find a fluxion of a given fluent and to find the original fluent from a given fluxion. The problem of finding quadratures is directly related to the problem of finding fluents from given fluxions.

One of the techniques to find quadratures and fluxions of curves is based on the binomial theorem. Newton developed a general version of the theorem in 1665. The theorem provides an expression for the expansion of $(1 + x)^n$. The expression is given by the series

$$1 + \frac{n}{1}x + \frac{n(n-1)}{1\cdot2}x^2 + \frac{n(n-1)(n-2)}{1\cdot2\cdot3}x^3 + \frac{n(n-1)(n-2)(n-3)}{1\cdot2\cdot3\cdot4}x^4 - \cdots.$$

The result had been known for the case in which n is an integer. Newton was able to obtain the expansion when n is a fractional number. Typical of Newton, he did not publish his work, but instead referred to results of it in his correspondence in the 1670s.[40] The first printed version of the general expression of the binomial theorem appeared in a 1688 tract by the physician Archibald Pitcairne, who attributed the result to the mathematician David Gregory, friend and fellow Scot.[41]

How Newton obtained the quadrature of a curve in the mid-1660s using the binomial theorem is illustrated with a particular curve within a family of curves of interest to Newton. The curve is given by $y = (1 - x^2)^{7/2}$ and is shown in the diagram.

On using Newton's binomial theorem, this curve can be expressed as

$$y = 1 - \frac{7}{2}x^2 + \frac{\frac{7}{2}\left(\frac{7}{2}-1\right)}{1\cdot 2}x^4 - \frac{\frac{7}{2}\left(\frac{7}{2}-1\right)\left(\frac{7}{2}-2\right)}{1\cdot 2\cdot 3}x^6 + \frac{\frac{7}{2}\left(\frac{7}{2}-1\right)\left(\frac{7}{2}-2\right)\left(\frac{7}{2}-3\right)}{1\cdot 2\cdot 3\cdot 4}x^8 - \cdots.$$

Prior to Newton's work, the English mathematician John Wallis was able to find the quadrature of simple curves of the form $y = x^m$, where m is a positive integer. The quadrature of this curve for values of between 0 and some number a is given by $a^{m+1}/(m+1)$.[42] In order to find quadratures for curves such as $y = (1 - x^2)^{7/2}$, it is necessary only to expand $(1 - x^2)^{7/2}$ using the binomial theorem and then to apply Wallis's result term by term to the infinite series.

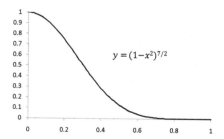

$$y = (1-x^2)^{7/2}$$

The binomial theorem can also be used to obtain fluxions, for example the fluxion of x^n. The modern approach to calculus would have us find

$$\lim_{\Delta t \to 0} \frac{\left[x(t+\Delta t)\right]^n - \left[x(t)\right]^n}{\Delta t},$$

where Δt is an increment of time. Newton's approach was similar but different. In an infinitely small period of time, which Newton denoted by o, the fluent x is augmented by its moment to become $x + o\dot{x}$ after that period of time. The fluxion of x is determined from

$$\frac{\left(x + o\dot{x}\right)^n - x^n}{o}.$$

On using the binomial theorem and after a little simplification, the fluxion can be written as

$$\frac{nx^{n-1}}{1}\times\frac{o\dot{x}}{o} + \frac{n(n-1)x^{n-2}}{2\cdot 1}\times\frac{o^2\dot{x}^2}{o} + \frac{n(n-1)(n-2)x^{n-3}}{3\cdot 2\cdot 1}\times\frac{o^3\dot{x}^3}{o} + \cdots.$$

When o is small but finite, then $o/o = 1$. After applying this result, then when "vanishes," the expression, and hence the fluxion, reduces to $nx^{n-1}\dot{x}$.

In his book *The Analyst*, the philosopher and Bishop of Cloyne George Berkeley pointed to logical problems with this approach.[43] The term o is assumed nonzero at the beginning of the proof and then set to zero to conclude the proof, Berkeley said. This is a logical fallacy: an assumption and its negation are both assumed in order to obtain the proof. The expression $nx^{n-1}\dot{x}$ for the fluxion is correct, but the mathematical foundations of the calculus behind obtaining it would not be properly established until the nineteenth century.

Another example of the use of the binomial theorem is the infinite power series expansion of the natural logarithm of a number $1 + a$. To use seventeenth- and eighteenth-century jargon, the hyperbolic logarithm of $1 + a$ is the area under the hyperbolic curve $y = (1 + x)^{-1}$ between 0 and the number a. Given that the binomial expansion of $(1 + x)^{-1}$ is $1 - x + x^2 - x^3 + \dots$, then the area under the hyperbola is $a - a^2/2 + a^3/3 - a^4/4 + \dots$ on using Wallis's result to find the area under the curve given by $y = x^m$. This infinite series expansion for the natural or hyperbolic logarithm of $1 + a$ was found by Nicholas Mercator in 1668, independent of Newton's binomial theorem.[44] Hyperbolic or natural logarithms is a prominent feature of some important areas of De Moivre's mathematical work.

The study of series expansions of various types became a staple of the British mathematical diet in the late seventeenth and early eighteenth centuries. This was not necessarily the case for the continental part of the Republic of Letters, as can be seen by comments from the mathematician and prominent member of the French Académie des sciences Pierre de Maupertuis. When James Stirling's *Methodus Differentialis* and Abraham De Moivre's *Miscellanea Analytica* were published in 1730, Maupertuis commented that "this business of series, the most disagreeable thing in mathematics, is no more than a game for the English; this book [Stirling's] and that of M. de Moivre are the proof."[45] These are books that contain a substantial amount of material on infinite series. The game that Maupertuis complained of was, however, substantial. Many of the more complex series expansions from the seventeenth and early eighteenth centuries may be found in Giovanni Ferraro's *The Rise and Development of the Theory of Series up to the Early 1820s*.[46]

In the first edition of *Doctrine of Chances*, published in 1718, De Moivre states that problems in probability could all be solved using the binomial theorem and infinite series.[47] The application of the binomial theorem to probability can be easily described. When n is a positive integer, the expansion of $(a + b)^n$ is expressed as

$$a^n + \frac{n}{1}a^{n-1}b + \frac{n\cdot(n-1)}{1\cdot 2}a^{n-2}b^2 + \frac{n\cdot(n-1)\cdot(n-2)}{1\cdot 2\cdot 3}a^{n-3}b^3 + \cdots + b^n,$$

using the binomial theorem. In that form it may be applied to problems in probability in three ways. (1) If a stands for success and b for failure, then the coefficient of the term $a^i b^{n-i}$ in the expansion is the number of ways of obtaining i successes and $n-i$ failures. (2) If a is the number of chances leading to success and b is the number leading to failure, then the number of chances of obtaining i successes and $n-i$ failures is the same coefficient as before times the number computed from $a^i b^{n-i}$. (3) If a is the probability of success and b the probability of failure, then the probability of obtaining i successes and $n-i$ failures is again the same coefficient times $a^i b^{n-i}$.

Prior to De Moivre's work in probability, there were two British mathematicians, one English and one Scottish, who used applications of the binomial theorem to solve probability problems. Both knew De Moivre and both were fellows of the Royal Society. The first is Francis Robartes—amateur scientist and mathematician, Whig politician, and younger son of the 1st Earl of Radnor. His work on the binomial appears in John Harris's *Lexicon Technicum* in an article entitled "Play."[48] The article is a loose translation of Christiaan Huygens's 1657 tract on probability, *De ratiociniis in ludo aleae*. Robartes used the binomial expansion to solve the classical problem of the division of stakes, one of the problems that led Blaise Pascal and Pierre de Fermat to the initial development of the probability calculus in 1654. There are also hints of the use of a binomial expansion in a paper on probability that Robartes presented to the Royal Society in 1692 but did not publish.[49] The other is John Arbuthnot—physician, mathematician, and satirical writer. He was also the friend and fellow countryman of Archibald Pitcairne and David Gregory. His use of the binomial expansion is part of his paper to the Royal Society on divine providence.[50] There Arbuthnot used the expansion to show that the probability of an equal number of male and female births in any year would be very small in a large population.

It was in 1687, the year when Abraham De Moivre arrived in London, that Newton published his magnum opus *Principia Mathematica*. Actually, it was Edmond Halley who brought the manuscript to press and the same Edmond Halley who encouraged Newton to write the manuscript. The Newton scholar Bernard Cohen has given an excellent and succinct summary of the *Principia*'s contents and its novelty.

> Newton's *Principia* is a remarkable book on many levels. It contains original results in pure mathematics (theory of limits and geometry of conic sections), it develops the primary concepts of dynamics (mass, momentum, force), it codifies the principles of dynamics (three laws of motion), and it shows the dynamical significance of Kepler's three laws of planetary motion and of Galileo's experimental conclusion that bodies with unequal weights will fall freely (at the same place on earth) with identical accelerations and speeds. It develops the laws of curved motions, the analysis of pendulums, and the nature of motions constrained to surfaces, and it shows how to deal with the motion of particles in continually varying force fields. Newton also indicates the way to analyze wave

motions, and he explores the manner in which bodies move in various resisting mediums. The crown of all appears in the final book 3, in which he discloses the Newtonian system of the universe—regulated by gravity, by the action of a general force, of which one particular manifestation is the familiar terrestrial weight. Here Newton treats at length of the orbits of planets and their satellites, the motions and paths of comets, and the production of tides in the sea.[51]

The exposition of mathematics in the *Principia* relied heavily on geometry, rather than on the newly emerging calculus. Further, the mathematics was complex enough that it could be read only by skilled mathematicians. The *Principia* went through two more editions during Newton's lifetime. Although the new generation of mathematicians in the early eighteenth century were familiar with the calculus, the *Principia* retained its geometrical approach and remained that way in England until the middle of the eighteenth century.[52] It was the geometry in it that first attracted De Moivre to the *Principia*.

The *Principia* had an enormous impact on science for well over a century after its publication. The ideas in the *Principia* were quickly adopted in England by scientists of the day. An English translation by Andrew Motte came out two years after Newton's death, and there were several publications that tried to explain the general ideas in the book. On the Continent, the major early adherents to Newtonianism were the Bernoulli brothers, Jacob and Johann, soon followed by their students and other members of the Bernoulli family. The twist on the Continent was that these mathematicians dropped Newton's geometrical approach and used the new calculus, the Leibnizian version in particular, to deal with the issues and topics raised in the *Principia*.[53] They also tended to side with Leibniz in the priority dispute that arose between Newton and Leibniz over the discovery of the calculus. Up to the first quarter of the eighteenth century, there was some resistance in France to the complete acceptance of Newtonian cosmology. Several prominent members of the French Académie royale des sciences continued to try to justify the Cartesian system of vortices over Newton's theory of gravitation to account for how celestial bodies travelled in their orbits.[54] By the late 1730s the *Principia* had made such inroads on the Continent that two Minim friars, Francis Jacquier and Thomas Le Seur, published their own edition of the *Principia* with commentary. The complete text of Newton's third edition was accompanied by a proposition-by-proposition commentary using Leibnizian calculus that was equal in length to the original text.

Akin to Darwin's *Origin of Species* in the nineteenth century, the *Principia* had an immediate impact on religious belief as well as on scientific enquiry. Through mathematics, Newton had demonstrated regularity and unity in the physical working of the cosmos. This supported a shift in the religious or philosophical thinking among several intellectuals, both clerical and lay. For some believers, the *Principia* strengthened the biblical role of God, the creator; God was the maker or designer of a well-ordered mechanistic universe whose operation could be predicted through mathematics. For others, God lost his biblical role as the king or shepherd of his people and was reduced only to the role of creator. The commonly used analogy was that God was a clockmaker who built his clock, got it running, and then left it to run on its own.[55]

The ideas in the *Principia* also impacted the interpretation of random events, the study of which, through games of chance, was one of Abraham De Moivre's major occupations. Newton had shown that the universe was subject to mathematical laws. Probabilists like De Moivre showed that random events also followed certain mathematical laws, although not in a deterministic way. The consequence of this insight is that the outcomes of random events are no longer subject to fate or fickle fortune.

❧ 3 ❧

Getting Established in England

Although De Moivre had left religious turmoil behind in France, he stepped into the midst of Protestant unrest in England. During the seventeenth century, England had its own problems with respect to religion and politics. There had been a civil war in the mid-seventeenth century which saw the king deposed and executed, and the Church of England disestablished as the state church. Upon the Restoration in 1660, the Church of England was again made the state church and other denominations that did not adhere to the established church were subjected to discriminatory acts of Parliament. When Charles II died in 1685, his brother James was proclaimed king. However, James II had openly converted to Catholicism. The situation was tolerated by the political and religious elites until James's wife gave birth to a boy, an heir who would be raised a Catholic. Fear of further religious and political upheaval in England spread; James was deposed in the Revolution of 1688 and replaced by his Protestant daughter Mary and her husband, William of Orange. James fled to France and, supported by Louis XIV, plotted his return to the throne, as did his son after James's death in 1701. One of the key players who brought William and Mary to the throne was William Cavendish, 1st Duke (at the time 4th Earl) of Devonshire. A Whig peer, Devonshire was one of a group of seven who had invited William of Orange to invade and was instrumental in the decisions of the Convention Parliament of 1689 that made William and Mary joint monarchs.

About two years after his arrival in England, De Moivre had an audience with Devonshire.[1] This interview was a significant one because Devonshire, a powerful Whig nobleman, was taking time to meet with De Moivre who, at the time, was a nobody. As such, De Moivre would not have just come in off the street to see Devonshire, but must have had a letter of introduction or a reference from someone else.

De Moivre's sponsor was likely Isaac Garnier, the apothecary to Charles II who had fled France in 1682.[2] In view of their professions in Vitry-le-François, Isaac Garnier as an apothecary and medical practitioner and Daniel De Moivre senior as a surgeon, the two Protestant families undoubtedly knew one another in France. Abraham De Moivre's connection to the Garnier family continued well beyond their common origins in Vitry-le-François. Garnier's elder son, Isaac Junior, married Eleanor Carpenter, the sister of Lord George Carpenter, 1st Baron Carpenter of Killaghy.[3] In 1738, De Moivre dedicated the second edition of *Doctrine of Chances* to the first baron's only child, Lord George Carpenter, 2nd Baron Carpenter. Isaac Garnier's younger son, Thomas, subscribed to De Moivre's *Miscellanea Analytica*, published in 1730.

Regardless of how the meeting was arranged between De Moivre and Devonshire, it helped to propel De Moivre forward in his future career as both tutor and scientist. One of the results of the meeting was that De Moivre was probably hired to tutor one of the duke's younger sons, James, as well as the elder son and heir, William, later 2nd Duke of Devonshire.[4] This was an early step in a long career of working as a mathematics tutor, and in a sense, followed in the footsteps of De Moivre's own tutor, Jacques Ozanam. However, there are two major differences between Ozanam's and De Moivre's teaching careers. While Ozanam taught children of the bourgeoisie, De Moivre typically taught the children of aristocrats and families with landed interests. Secondly, while Ozanam died nearly penniless, De Moivre died comfortably holding £1600 in capital invested in annuities.[5] The second and unexpected result of the meeting with Devonshire was De Moivre's first brush with Britain's scientific community. Late in life, De Moivre told his friend and biographer Matthew Maty that as he approached the duke's house, he saw a man whom he did not know come out of the house. De Moivre later discovered that the man was Isaac Newton, with whom he eventually became close friends. While waiting in an anteroom for the duke, De Moivre noticed a copy of Newton's *Principia Mathematica* and glanced through it. As Maty writes,

> The illustrations it contained [several geometrical diagrams] led him to believe that he would have no difficulty reading it. His pride was greatly injured, however, when he realized that he could make neither head nor tail of what he has just read, and rather than propel him to the forefront of science, as he had anticipated, his studies as a young scholar had merely qualified him for a new development in his career. He rushed out to buy the *Principia*, and as the need to teach mathematics as well as the long walks he was thus forced to take around London left him scarce free time, he would tear out pages from the book and carry them around in his pockets so that he could read them during the intervals between the lessons.[6]

In France, De Moivre had been well trained in geometry and Cartesian physics. Therein, the standard Cartesian text, Rohault's *Traité de physique*, was relatively nonmathematical in contrast to Newton's *Principia*, which was filled with geometrical arguments and diagrams.

On December 16, 1687, very soon after their arrival in London, the Moivre brothers, Abraham and Daniel, were made denizens of England.[7] This legal status gave them the ability to buy land in their new country and came at a monetary cost of about £25—indicating the brothers were far from destitute when they arrived in England.[8] A few years later in 1706, Abraham, but not his brother, became a full citizen of England.[9] This also came at a cost of about £65, which was more than the annual salary of £60 that a well-placed clergyman would make.[10] As part of becoming a naturalized Englishman, De Moivre had to receive the sacrament of Holy Communion in the Church of England. De Moivre, with two of his Huguenot friends, Gideon Nautanie and John Mauries, as well as many other Huguenots, received the sacrament on December 9, 1705, at St. Martin-in-the-Fields Church. The three new citizens each in turn attested to the other two taking Communion at the church.[11]

In the 1690s De Moivre tried to regularize his teaching by obtaining a permanent teaching position. Two Royal Academies were proposed in 1695, one to be situated in Covent Garden near where De Moivre lived. Instruction in the academies would be given in a variety of subjects including languages, mathematics, and music. De Moivre was listed as one of the two mathematics teachers and his brother Daniel as a flute teacher. Expecting a significant response to the opening of the academies, entrance was by lottery. There were 40,000 tickets, sold at a price of £1 each, from which 2,000 winners would be drawn. The winners were to gain entrance to the academies and could take the subject of their choice. For a variety of reasons, the lottery scheme fell apart and the school masters themselves opened the academies to general subscription.[12] This was also unsuccessful and the entire project failed. Abraham De Moivre continued to work as a private tutor in mathematics and his brother continued teaching the flute privately.

The political sympathies of Huguenot refugees generally lay in the direction of the Whig faction,[13] and Devonshire was a leading Whig politician. In light of the failed attempt to regularize his teaching career, De Moivre soon expanded his tutoring clientele, probably through Devonshire's connections. Indeed, many of De Moivre's students were from leading Whig families. One very prominent Whig, closely connected politically to the Duke of Devonshire, was Robert Walpole, who is considered to be Britain's first prime minister. It is likely that De Moivre tutored Walpole's eldest son, Edward. Moreover, while he was in power, Robert Walpole ran an enormous web of political patronage. Many of the subscribers to De Moivre's 1730 *Miscellanea Analytica* were part of Walpole's web; many of the same subscribers were tutored by De Moivre.[14] Another possible source of students for De Moivre via Devonshire was through Devonshire's own family ties. Devonshire's distant cousin Elizabeth Cavendish married Ralph Montagu, 1st Duke of Montagu. In 1706 Montagu hired De Moivre to tutor his son John.[15] Over time, De Moivre tutored several members of the extended Montagu family, including the amateur mathematician and Member of Parliament Edward Montagu. Although a good mathematician in his day, he is mainly known today for being the husband of author and literary hostess Elizabeth Montagu, "Queen of the Bluestockings."[16]

As noted already, the typical social background of the students that De Moivre taught runs a narrow gamut from the aristocracy and their relatives to the landed gentry and families acquiring landed status. Despite this general trend, De Moivre did not find his students exclusively from among the sons of the elite. Some of his students came from the Huguenot refugee community or were recommended to him by other Huguenot tutors living in England. One of his students from the very early 1690s was Peter de Magneville. In the latter part of the 1690s Magneville was working or studying with Johann Bernoulli, who at the time held the chair of mathematics at the University of Groningen in the Netherlands. Later, after his brother Jacob's death in 1705, Bernoulli took his brother's professorship in mathematics at the University of Basel. Other Huguenot students include Michel de la Roche and Peter Davall, who became a journalist and a barrister, respectively. Other Huguenots who were likely students of De Moivre include Francis Fauquière, son of Newton's deputy at the Mint, and Isaac Guion, son of a Huguenot distiller of the same name. Martin Folkes, who was not of Huguenot descent, studied with De Moivre in his youth. Folkes's original tutor was Jacques Cappel, formerly a teacher at Saumur. It is probable that De Moivre came to tutor Folkes through Cappel's recommendation.

Many of De Moivre's tutor-student relationships turned into life-long friendships. For example, in his will Peter de Magneville left £20 to De Moivre, as well as another £20 to Michel de la Roche, "in consideration of our antient friendship."[17] Martin Folkes and Edward Montagu, both tutored by De Moivre prior to 1710, dined with their former teacher on his eightieth birthday in 1747.[18] The same Montagu showed solicitousness and other signs of close friendship. Writing to his wife at their country home in 1751, Montagu asked his wife to send wheatears, an avian delicacy, to De Moivre, care of Pons Coffeehouse, because he thought De Moivre might like to see how they taste.[19] After Montagu became ill in the late 1760s, he presented the only known portrait in oil of De Moivre to the Royal Society.[20] Presumably, Montagu also commissioned the portrait from Joseph Highmore who painted it in 1736.

De Moivre must have possessed a strong charisma to maintain such good friendships for thirty, forty, and even fifty years. In fact, hints of it come from travelers who met De Moivre and wrote about their experiences. Charles-Étienne Jordan wrote in 1735 that when he met De Moivre in London two years earlier during his travels, De Moivre was pleasant company and also man of wit.[21] More than a decade later in 1747, the Huguenot clergyman Jean Des Champs arrived in London to settle there. Living only a third of a mile from De Moivre's lodgings, he met and dined with De Moivre, thirty years his senior, several times soon after his arrival. He described De Moivre as *très joyeuse compagnie*, which a translator has rendered, in an understated way, "good company."[22] In his biography of De Moivre, Matthew Maty adds, "Strength and depth rather than charm and liveliness were the hallmarks of his conversation and writing."[23]

As De Moivre cultivated his connections with the Whig elite in order to further his career as a tutor, he also developed friendships within the French émigré community. The Huguenot literary intelligentsia met regularly for coffee and conversation at the Rainbow Coffeehouse near St. Martin-in-the-Fields Church.[24] Led by Pierre Des Maizeaux, other early members of the group included Abraham De Moivre, Pierre Coste, Pierre-Antoine Motteux, and Peter Davall. Despite history viewing De Moivre mainly as a mathematician, De Moivre was well acquainted with classical French literature, including the works of the poet and fabulist Jean de La Fontaine and of the writer of fantasy and satire François Rabelais. He also enjoyed the plays of Pierre Corneille and Molière (Jean-Baptiste Poquelin) and had read the *Essais* of Michel de Montaigne. He had seen Molière's *Le Misanthrope* performed when he was a young man in France and could recite the lines of the play by heart. His favorites among these authors were Rabelais and Molière. Over the years 1693–1694, Motteux revised and completed a translation of Rabelais' *Gargantua and Pantagruel* begun 30 years before by Sir Thomas Urquhart. One can imagine some of the discussion at the Rainbow Coffeehouse about what Motteux had done. Book V, the last book of *Gargantua and Pantagruel*, was thought by many to be spurious. However, De Moivre not only was convinced that Book V was true Rabelais, but also thought it the best part of the entire work.[25]

In his early years in England, De Moivre interacted with some of the established mathematicians. He met Nicholas Fatio de Duillier in about 1692.[26] Of Swiss origin, Fatio had come to England in 1687 and for a time had a close friendship with Newton that eventually ended with a falling out. De Moivre's relationship with Fatio is fraught with contradictory claims as to the nature of their relationship. One claim holds that De Moivre learned Newton's new calculus from Fatio,[27] while the other, originating with an anecdote of Magneville, argues that Fatio received tutoring in mathematics from De Moivre.[28]

Whatever their teacher-student relationship, they became competitors in a mathematical problem on cycloids proposed in 1697 by Jacob Bernoulli via his brother Johann. The problem is related to the brachistochrone (or curve of quickest descent) problem that Johann Bernoulli had posed (and solved) the year before as a challenge to the mathematical community. Four of the great mathematicians of the day rose to the challenge and also solved Johann Bernoulli's brachistochrone problem: Newton, Leibniz, l'Hôpital, and Jacob Bernoulli. In a letter to Henri Basnage de Beauval, a prominent Huguenot exile living in Rotterdam, Johann Bernoulli reviewed the solutions to the brachistochrone problem in a general way and then posed his brother Jacob's problem on cycloids.[29] The letter was published in Basnage's *Histoire des ouvrages des sçavans*, a highly reputable journal out of Rotterdam that published a wide variety of mainly nonmathematical articles, including ones by leading figures in the Republic of Letters such as Christiaan Huygens and Gottfried Leibniz. Both De Moivre and Fatio de Duillier found the problem while reading *Histoire des ouvrages des sçavans*. They worked intensely on the problem for some time, but neither had any success in finding a solution.[30]

De Moivre met the astronomer Edmond Halley in about 1692.[31] Halley was acquainted with Huguenot merchants in London through the Levant Company.[32] He had also been to Saumur in 1681 as part of his own scientific version of the Grand Tour on the Continent. During his tour, rather than participating in the usual cultural attractions, Halley visited several astronomers in France and Italy.[33] At Saumur, Halley would have met De Moivre's teachers rather than De Moivre himself since De Moivre did not attend the academy until 1682. It is probable that Halley and De Moivre met through mutual Huguenot acquaintances in London—either through Levant company connections or a former teacher from Saumur, such as Abraham Meure or Jacques Cappel.

Halley often discussed mathematical problems with De Moivre, Halley having an application in need of a mathematical solution and De Moivre seemingly interested in a challenging mathematical problem to work on. Their different interests reflect a classification of mathematics prevalent in the eighteenth century that divided the subject into pure and mixed mathematics.[34] The dividing line, as described in eighteenth-century encyclopedias,[35] is based on how quantity is treated. In pure mathematics, quantity is considered "abstractly, and without any relation to matter or bodies." In mixed mathematics, quantity is considered "as subsisting in material being; e.g. length in a road, breadth in a river, height in a star, or the like." Halley was a very capable mathematician with interests mainly in mixed mathematics such as astronomy. At least early in his career, De Moivre could be described as a pure mathematician in the eighteenth-century meaning of the term.

Rubens's stereographic projection (from the Getty Research Institute collection).

On at least three occasions, Halley came to De Moivre for his thoughts on some mathematical problems. The first, motivated by issues concerning navigation at sea, was a problem in spherical geometry arising from projecting a sphere onto a plane (a stereographic projection). The problem can be explained by the engraving of a stereographic projection designed by Peter Paul Rubens and published in 1613.[36] The lines of longitude and latitude in the sphere held by Atlas are projected by the light of the putto's torch onto circles on the ground. What De Moivre essentially told Halley in their discussions late in 1695 or early 1696 was that the angles at which the circles on the ground intersect are the same as the associated spherical angles on the sphere obtained by two intersecting circles on the sphere. Halley went away and proved the result on his own.[37] The second recorded mathematical discussion arose in 1700 from a problem concerning refracted light in rain showers resulting in rainbows. De Moivre showed Halley how to obtain the ratio of refraction for rainwater by measuring the diameter of a secondary rainbow in the sky. The ratio of refraction is the ratio of the sine of the angle of incidence of a ray of light to the sine of the angle of refraction of the ray through the medium, in this case rainwater.[38] The ratio of refraction for a given medium is constant at all angles of incidence. The third discussion occurred in 1706 when Halley was involved in a publication of mathematical tables and some related essays.[39] His part was an essay on the valuation of fixed-term annuities. De Moivre verified a formula Halley had obtained to determine the approximate interest rate when the value and term of the annuity were given.[40] The approximation was simple as well as accurate; approximations that had been obtained in the 1670s required a substantial amount of calculation. All these interactions have a connection to areas of mathematics in which De Moivre was working—geometry that he had learned in France and his newly emerging interests in power series and solutions to polynomial equations. The determination of the ratio of refraction required the solution to a biquadratic equation (a fourth-degree polynomial with no odd powers). The determination of the interest rate required a power series approximation (involving powers of the interest rate that was to be determined) followed by the solution to a quadratic equation.

Soon after meeting Halley, De Moivre finally met the man whom he saw coming out of the Duke of Devonshire's residence in 1689. Newton and De Moivre became good friends. Probably after Newton took up residence in St. Martin's Street in 1711, they would often meet at Slaughter's Coffeehouse in St. Martin's Lane, after De Moivre had finished teaching for the day. They would then move to Newton's house, which was close by, for an evening of conversation and philosophical debate. De Moivre came to know Newton well enough that Newton told De Moivre several details about his early life.[41]

When Halley and De Moivre first met, Halley was assistant secretary to the Royal Society. It was three years after their initial meeting that Halley brought some of De Moivre's work to the attention of the Royal Society. In the minutes for the meeting of June 26, 1695, there is an entry stating:

Halley related that one Mr Moivre a French Gentleman had lately discovered
to him an Improvement of the Method of Fluxions or Differentialls invented
by Mr. Newton with a ready application thereof to Rectifying of Curve lines,
Squaring them and their Curve Surfaces, and finding their Centers of Gravity
&c.[42]

Since his first glance at it in 1689, De Moivre had read the *Principia* thoroughly.
The idea for the paper that he gave to Halley, which was published in *Philosophical
Transactions*,[43] came out of a result in the second book of the *Principia* regarding the
fluxions of products, quotients, and powers of variables.[44] De Moivre was interested
in the other side of the calculus—finding an integral, to use modern terminology.
This operation yields areas under curves (quadrature), lengths of curves (rectification), and centers of gravity or balance points of curves. Using some very astute
mathematical tricks, De Moivre expanded the types of curves for which integrals or
inverse fluxions could be found. This was well beyond the curves that Newton had
treated in this part of the *Principia*.

After reading his copy of *Philosophical Transactions*, the doyen mathematician of the day, John Wallis, Savilian Professor of Geometry at Oxford, wrote to
the secretary of the Royal Society. While suggesting that the Royal Society publish
some early letters of Newton that "are more to the purpose than that of De Moivre,"
Wallis commented, "Who this De Moivre is, I know not."[45] This obscure foreign
mathematician was soon to be much less obscure in Royal Society circles.

On June 16, 1697, the second of De Moivre's papers was read before a meeting
of the Royal Society. The subject matter once again dealt with Newton's mathematics.
This time it was an extension of Newton's binomial theorem to the multinomial case.
Where Newton had essentially obtained an infinite series expansion for curves of the
form $y = (a + bx)^n$, De Moivre obtained a series expansion for the curve $y = (ax +
bx^2 + cx^3 + dx^4 + ...)^n$.[46] Election to fellowship in the Royal Society followed five
months later on November 30, 1697, and his scientific career was well underway.[47]
De Moivre continued to work on a topic related to the multinomial. Soon after his
election, he presented a paper that showed the reader how to invert a series. Given
the equation $py + qy^2 + ry^3 + sy^4 + ... = ax + bx^2 + cx^3 + dx^4 + ...$, De Moivre was
able to solve for y alone as a power series in x.[48]

In 1700, he, in a sense, returned to his youth by taking up a classical problem in
geometry. In this work he studied the resulting solid shapes obtained when segments
of the Lune of Hippocrates are rotated about an axis. A lune is the crescent-shaped
figure obtained by overlapping two circles and removing the area of the larger circle.
The Lune of Hippocrates, first studied by Hippocrates of Chios in the fifth century
BCE, is obtained when one circle is twice the area of the other and the circumference
of the smaller circle passes through the center of the larger one. De Moivre's 1700
paper may have been motivated by a letter from John Wallis published the previous
year in *Philosophical Transactions*.[49] It was about the areas under parts of the Lune

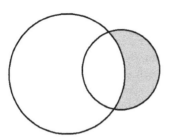

A simple lune.

of Hippocrates. Wallis read De Moivre's paper, as well as the entire issue, the day he received his copy of *Philosophical Transactions*. In a letter to Hans Sloane, secretary of the Royal Society, Wallis wrote something that any reader of De Moivre's work finds at time frustratingly typical: De Moivre had merely stated his results and made no effort to demonstrate the validity of them. Wallis suggested to Sloane that De Moivre's results could be verified using a result in *Mechanica*, a book Wallis had published in 1670.[50]

By 1700 De Moivre had made a good start to his scientific career. At about the same time, he took some small steps into the wider Republic of Letters, one in the arts and one in the sciences. The arts side began with the arrival in England of Abbé Jean-Baptiste Dubos in 1698, about a year after the end of the Nine Years' War, also known as the War of the Grand Alliance or King William's War. Initially trained in theology, Dubos was working as a diplomat representing Louis XIV. He met De Moivre sometime before his return to France in about 1702. The scientific step was meeting Jacob Hermann, who had studied mathematics with Jacob Bernoulli in Basel. During 1701 and 1702 Hermann visited Holland, England and France to meet members of the scientific community.[51] In England he hoped to meet Newton; he met De Moivre instead. On August 1, 1701, Hermann wrote from Paris to De Moivre's friend Pierre Des Maizeaux.[52] The letter contains some mathematical calculations related to a problem in physics. At the end of the letter, Hermann added a request that Des Maizeaux pass on his compliments to De Moivre and to other good friends.

Both steps into the Republic of Letters were short and short-lived. Soon after his return to France, Dubos retired from public life and devoted himself to his studies, writing on history and on a theory of painting and poetry. There seems to have been no further contact between De Moivre and Dubos until about thirty-five years later when another visitor from France, art critic Abbé Jean-Bernard Le Blanc, met De Moivre. After Le Blanc conversed with De Moivre about Dubos's book *Reflexions critiques sur la poesie et sur la peinture*, De Moivre sent his compliments to Dubos via Le Blanc and asked for a copy of a forthcoming new edition of the book. De Moivre's contact with Hermann lasted a little longer. In 1706 he wrote to Hermann in Basel, sending his letter through Johann Bernoulli, who was now in Basel. He received a reply, but there is no evidence of any further correspondence, although De Moivre is mentioned several times between 1702 and 1718 in correspondence between Hermann and Johann Bernoulli. The main gist of the 1706 letter, which is not extant, seems to have been a report on a result obtained by a young unnamed friend of De Moivre.[53] It was a series approximation to the number π that converges rapidly. The series correctly gives π to four decimal places (3.1416) after three terms and to six decimal places (3.141593) after five terms.

A brief look at *Philosophical Transactions* shows that publications flowed from someone like Edmond Halley; from De Moivre, it was more of a trickle—thirteen from De Moivre compared to more than fifty from Halley over their lifetimes. Further, nearly one third of De Moivre's output of scientific papers occurred during the first five years of his scientific career. One reason for the difference can be attributed to De Moivre's need to tutor during the day in order to make a living. Another reason is described by Maty:

> I learned that it had been Mr. De Moivre's preference from the very outset to work on these difficult problems at night rather than in the day, since they required a great deal of attention; and that, several years later, whenever he felt able to fix his mind on the most complex calculations even during the day, he could not tolerate noise in the house, as the disturbance upset his concentration.[54]

De Moivre lived in or near St. Martin's Lane. During his lifetime, the east side of the lane housed traders and artisans while the west side was a fashionable location for doctors and artists.[55] It could have been a noisy area during the day, particularly if there was significant horse and carriage traffic on a cobbled street.

In addition to introducing De Moivre to scientific society, Halley may have helped to establish De Moivre's teaching career. Halley had connections to Christ's Hospital, a charitable school originally located in London, now in Horsham, West Sussex. Halley lived near the school, had been a mathematical examiner there, and had consulted the school's registers for information on youth mortality while he was preparing his 1693 paper on the analysis of the bills of mortality from Breslau.[56] Possibly through Halley's recommendation, De Moivre was appointed a mathematical examiner at Christ's Hospital in 1698, three years after Halley last served in that position. The job paid £2 for a day or two of work. De Moivre continued to examine students until 1702. The following year he was succeeded by John Harris, author of *Lexicon Technicum*.[57] De Moivre may have either given up or lost this job due to illness; by the latter part of 1702 he had contracted smallpox and was recovering from it by early December.[58]

While Abraham was solidifying his position as a tutor, as well as an able mathematician within the scientific community, his brother Daniel was developing his career as a talented flautist.[59] Daniel would have played a Baroque flute, whose wooden construction in the body is more in line with a recorder, but still played transversely like a modern flute.[60] By the time the brothers had arrived in England, the flute was a favorite instrument among amateur musicians, ladies and gentle-man playing for their own amuse-ment and pleasure. Daniel catered to this group as a teacher, performer, and composer. In 1701, he published his first set of pieces for solo flute, a collection written particularly for stu-dents. The pieces were mainly writ-ten in dance form: allemandes, cou-rants, jigs, minuets, and sarabands. A second collection came out in 1704 and a third in 1715.[61] All the collec-tions were published by the prom-inent London music publisher John Walsh, whose shop in the Strand was a short walk from where both Dan-iel and Abraham lived. The latter two collections of flute music may have been of pieces composed for Dan-iel's own performances. While the 1701 music has the title *Lessons for a Single Flute*, the later publications do not have the word lessons in their titles. De Moivre's music was popu-lar; Walsh republished the three col-lections at various times in the 1720s and early 1730s. The only known recorded notice of one of Daniel's concerts is in 1717 when a subscrip-tion series of twelve concerts were performed by De Moivre and several others at Leathersellers' Hall.[62] Un-like his publisher's shop, the concert venue was some distance from home, a walk of about two and a quarter miles.

Until about 1700, or shortly thereafter, the Moivre brothers' ca-reers ran in parallel—successful but not outstanding in their chosen fields.

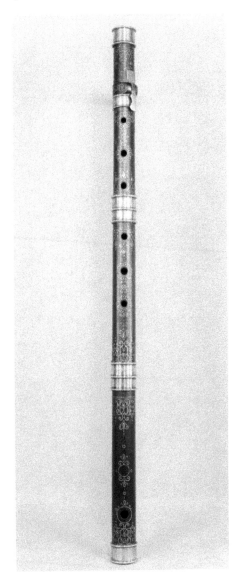

Baroque flute, circa 1700. (Photo © Victoria and Albert Museum, London.)

49

Soon Abraham De Moivre's mathematical career would accelerate well beyond his brother's musical one. What sparked the change in trajectory was an unpleasant "academic fight" that began to erupt in 1702.

❧ 4 ❧

Scotica Mathematica

George Cheyne was a Scottish physician who had studied medicine with Archibald Pitcairne in Edinburgh. With Pitcairne's sponsorship, Cheyne obtained his medical degree from King's College of the University of Aberdeen. Like many other Scots, he moved south to London where, along with fellow physician and mathematician John Arbuthnot, he became part of a circle of friends and former students of Archibald Pitcairne and David Gregory.[1] Adept at mathematics and initially unable to find employment as a physician in London, Cheyne became, like De Moivre, a teacher of mathematics to the upper classes. Unlike De Moivre, who tutored many students, Cheyne taught only one student, the younger brother of the Earl (later Duke) of Roxburghe.

Soon after his arrival in London in 1702, Cheyne became involved in an "academic fight" with De Moivre over the contents of a book that Cheyne had published entitled *Fluxionum Methodus Inversa*.[2] As the Latin title suggests, it was a book about inverse fluxions, or, in modern terms, the calculus of integration.[3] Later interpretations of the dispute have reflected negatively on De Moivre, describing him as the person who fronted some of Newton's dirty work. Certainly, this was the interpretation of the French mathematician and academician at the Académie royale des sciences Pierre Varignon, who later became one of De Moivre's continental correspondents in the Republic of Letters. Writing to Johann Bernoulli on May 2, 1711, about seven years after the height of the dispute, Varignon said that he had heard from a young Scottish physician visiting Paris that it was Newton who had provoked De Moivre to react to Cheyne's work.[4] On the other hand, Astronomer Royal John Flamsteed met De Moivre several times while the episode was underway during 1703 and was more sympathetic to De Moivre than what might be taken from Varignon's comment.

While he was still in Scotland, Cheyne wrote the manuscript for his book *Fluxionum Methodus Inversa*. Rather than containing new research on the topic of quadratures or on the finding of areas under curves, it was a survey of results that had been obtained by both British and continental mathematicians. What was new about the manuscript was that work on the subject up to that date was now systematized in one place. Cheyne put together a set of rules for obtaining quadratures and provided his own proofs of the results. Through Pitcairne, also in Scotland, the manuscript was sent to John Arbuthnot in London who took it to Newton. Cheyne wanted to know if the manuscript was worth publishing. Newton glanced at the manuscript and was lukewarm. Prior to or despite receiving Newton's opinion, the manuscript was sent to a printer by February of 1702. Already Cheyne's Scottish friends in London were championing it to the scientific community.[5] Throughout 1702 there were continual delays in printing. Joseph Raphson was recruited in May to proofread the book as the sheets slowly came off the press. By July, Cheyne's Scottish friends were silent and there were rumors that something was wrong with the book. After Cheyne himself arrived in London near the end of 1702, he was taken by Arbuthnot to meet Newton. During the meeting there was an apparent misunderstanding. According to the recollection of John Conduitt,[6] husband of Newton's niece, Newton thought Cheyne wanted money to get the book printed and offered it. Cheyne wanted Newton to peruse the manuscript and offer advice on the contents; he was offended by the offer of money. This recollection does not quite ring true since, by that time, the manuscript had been with the printer for several months and part of the book was already printed. Further, why would Newton offer money to have a book printed when he was lukewarm about the manuscript? With or without any offer of money from Newton, the printing of the book was finally completed in February of 1703, at which time Cheyne presented a copy to the Royal Society at one of its meetings.[7]

Newton was upset by the publication of Cheyne's book. For several years he had sat on his own results related to quadrature, allowing only a chosen few to see his work. With Cheyne's book in print, he felt that his work was threatened. Results were now being published that duplicated his unpublished work. Newton hinted in his *Opticks*, published in 1704, that his work on the quadrature of curves had been plagiarized.[8] If Varignon's interpretation of the affair is correct, then during 1703 Newton encouraged his friend De Moivre to write a reply to Cheyne's book in order to denigrate it.

Still seeking advice, Cheyne sent a copy of his book, shortly after it was published, through an intermediary to Johann Bernoulli in Groningen. After some prodding, Bernoulli sent Cheyne a list of corrections, as well as some rules for quadrature that he thought were easier to use.[9] On receiving this, Cheyne added a section to his book entitled, "Addenda & Adnotanda." Although there is a reference to Bernoulli at the end of the section, there is no hint that Bernoulli was mainly responsible for the section. Bernoulli was not happy with the added section. He felt that his comments had been corrupted in several places and that Cheyne's rendering of the material made Bernoulli look bad in the eyes of his colleagues.[10]

On reading John Flamsteed's letters to his friend Abraham Sharp, a slightly different picture emerges of the role that De Moivre played in the affair. Flamsteed met De Moivre in early December of 1702. De Moivre had seen Cheyne's manuscript prior to the printed book appearing, and was concerned by the contents of it. Flamsteed related the encounter to his friend Abraham Sharp in his somewhat idiosyncratic English:

> He [De Moivre] told me hee had seen Dr Sheens book and was mightly concerned at something in it. he threatned to explaine the subject of it now very planely and to remarke severall things in it that were amiss. You may be sure I did not discorage him from doeing a thin that may be for the advantage of science: he says he will write in English. I advised him to doe it in French rather and get a freind that was by and understands both the subject and that language to translate it for him which Course I hope hee will take and then wee shall have this business seached [searched] to the bottom.[11]

There is no mention of Newton in the letter. Over the course of 1703 and into 1704, several letters between Flamsteed and Sharp mention the dispute. Had there been any hint that Newton was behind De Moivre's criticism of Cheyne, Flamsteed likely would have said something to Sharp. When Newton became president of the Royal Society in 1703, Flamsteed was soon critical of him in a letter to Sharp.[12] Flamsteed pulled no punches when writing to Sharp about other people he did not like, such as Edmond Halley. When John Wallis died in 1703 and the Savilian Professorship of Geometry at Oxford became vacant, Flamsteed wrote to Sharp, "Dr Wallis is dead Mr Halley expects his place who now talkes sweares and drinks brandy like a sea captaine so that I much fear his own ill behaviour will deprive him of the advantage of this vacancy." Flamsteed also praised where he thought praise was due, commenting to Sharp about De Moivre that "he is an honester and abler man than any of them."

My own take on the matter is that a copy of Cheyne's as yet unpublished manuscript was circulating in the scientific community. De Moivre read the manuscript on his own, was concerned by its contents, and probably expressed his concerns to Newton as he had done to Flamsteed. The initiative to publish a reply was De Moivre's rather than Newton's, although Newton may have encouraged De Moivre once his decision was made.

De Moivre was actually well placed to comment on Cheyne's book. He had become something of an expert on the subject of quadratures; in June of 1702 De Moivre presented a paper to a meeting of the Royal Society on this topic.[13] It was published that year in the March/April issue of *Philosophical Transactions*; Flamsteed received his copy in July.[14] Using some series expansions, De Moivre found quadratures for curves of the form $y = x^n(ax \pm x^2)^{\pm 1/2}$ and $y = x^n(a^2 \pm x^2)^{\pm 1/2}$. Prior to its publication, De Moivre showed the paper to Newton. What becomes apparent from his action is that De Moivre is one of the chosen few in Newton's eyes. Once Newton had seen De Moivre's manuscript, he showed De Moivre his own manuscripts where he had obtained related results. This is one of the rare instances of Newton showing anyone the details of his unpublished mathematical work.

By the end of March 1704, De Moivre published his reply to Cheyne under the title *Animadversiones in G. Cheynaei Tractatum de Fluxionum Methodo Inversa*.[15] The book is in Latin, as is Cheyne's, rather than English or French. This was to the liking of Abraham Sharp, who had expressed in a letter to Flamsteed early in 1703 that he would not be able to read what De Moivre was going to write if it were in French.

Anita Guerrini neatly summarizes the contents of the *Animadversiones* by writing that De Moivre

> accused Cheyne of misunderstanding Newton's method. He only showed the method, said De Moivre, but did not derive it. De Moivre then enumerated in devastating detail the "many errors" in Cheyne's book.

This does not address the reasons why De Moivre was "mightly concerned" about Cheyne's book late in 1702. In the same letter to Flamsteed, Sharp gives a foreshadowing of some of these reasons. Sharp had read De Moivre's 1702 paper on quadrature in *Philosophical Transactions*. He found it "writ … ingeniously," but also noted that De Moivre had "left some particulars short which I have long desired to see further cleared." This echoes Wallis in comments on De Moivre's 1700 paper on the Lune of Hippocrates. Sharp was looking forward to seeing both publications to learn more about the topic. After De Moivre published his book, he gave a copy of it to Flamsteed. Not particularly interested in the subject of quadrature, Flamsteed gave his copy to Sharp.[16]

It is De Moivre's 1702 paper on quadrature that is a key to De Moivre's agitation, not necessarily Newton's concerns about plagiarism or other issues. In his *Animadversiones*, after an initial discussion, De Moivre goes directly to the point. On page 13 of his book, De Moivre picks a particular result from page 18 of Cheyne's *Fluxionum Methodus Inversa* and shows how the result can be obtained correctly using his 1702 paper. The paper is so central to some of his arguments that it is reproduced in full, except for the last two paragraphs, beginning on page 106 of the *Animadversiones*.

The road to publication for De Moivre was a little bumpy. His intention was to publish a letter in *Philosophical Transactions* pointing out the problems in Cheyne's work, but the Scottish faction within the Royal Society became a roadblock. The Scots within the Royal Society had been promoting Cheyne's work and they did not easily let go. Probably out of frustration, at one point De Moivre began examining some earlier Scottish work on quadrature and rectification of curves in order to criticize it, including David Gregory's *Exercitatio geometrica de dimensione figurarum* published in 1684. De Moivre thought he had found a problem with the book and told Flamsteed of it. For his own part, Flamsteed thought there were problems with Gregory's more recent book on astronomy. A few months after this exchange, Flamsteed claimed in a letter to Abraham Sharp that Gregory had put pressure on Halley who then went to Hans Sloane, the secretary of the Royal Society in charge of printing *Philosophical Transactions*. As Flamsteed put it, Sloane would not "permit any thinge to be printed that may reflect on any of his confederates."

The Scottish faction's interference may also explain another insert in De Moivre's book. The mathematician John Colson read a paper to the Royal Society in August of 1703 entitled "Methodus Universalis and [*sic*] Exposito pro Solutione equationum Analyscarum."[17] It was ordered to be printed but was never published. A letter from Colson to De Moivre dated November 15, 1703, appears on pages 53 through 67 of the *Animadversiones*. The letter is concerned with methods of solution (y in terms of x) of equations such as $y + axy + a^2y = x^3 + 2a^3$ where a is a known constant. Very likely the letter is related to what had been read before the Royal Society a few months earlier. Since the equation in x and y also defines a curve, Colson's method can be used to find the quadrature of the curve: express y as a power series in x and obtain quadratures term by term as Newton had done in the 1660s.[18] Near the end of the letter, Colson says that Cheyne was using a similar method of solution to these types of equations but had stopped when the problems became too difficult for him. At the end of the letter, Colson praises De Moivre for revealing and correcting the errors in Cheyne's book. Perhaps Colson criticized Cheyne in the paper he submitted to the Royal Society, and for this reason it was not published in *Philosophical Transactions*.

At the time, Colson and De Moivre were very good friends.[19] This was about a decade before Colson became a fellow of the Royal Society, so they must have met somewhere other than Royal Society meetings. They remained friends well into the 1730s. Colson was a subscriber to De Moivre's 1730 *Miscellanea Analytica*. In his 1736 translation and extensive gloss of a seventeenth-century Newton manuscript on fluxions, Colson referred to De Moivre as "my good friend" and "my ingenious friend."[20] This is the same Colson who became De Moivre's successful rival for the Lucasian Professorship of Mathematics at Cambridge in 1739.

The effect of Cheyne's book on De Moivre's scientific career and De Moivre's response was far-reaching, in the sense that it brought him into contact with continental mathematicians. Once his *Animadversiones* was published, De Moivre sent a copy of it on April 22, 1704, to Johann Bernoulli in Groningen and asked for comments. This is the first of an exchange of letters between De Moivre and Bernoulli that lasted a decade.[21] It is De Moivre's real debut into the wider Republic of Letters. Bernoulli made some notes on the *Animadversiones* and wrote back on November 15 with some comments. He felt that De Moivre had gone too far in some of his criticisms of Cheyne, that he passed over the errors in the Addenda & Adnotanda (De Moivre may not have seen this addition), and that De Moivre had made some mistakes of his own. On the last point, Bernoulli said that he may have been mistaken and was willing to send De Moivre the notes he had made. One final point that Bernoulli made (actually it was the first on his list) was that De Moivre's writing style was too prickly. It might lead the reader to think that De Moivre had other motives for writing the book other than his pursuit of the truth. Bernoulli's comment here is a reference to the mode of discourse in the Republic of Letters. Pursuit of truth was an ideal of the Republic of Letters and one of the unwritten rules was civility in discourse.[22] Bernoulli's implied advice to De Moivre is that his barbed comments may have crossed the line. In later publications, De Moivre took this advice to heart and expressed his severe criticisms of others through alternative venues.

Politically, the Pitcairne–Gregory group of Scotsmen were Tories, as opposed to De Moivre whose contacts were mostly Whigs. With Anne recently on the throne of England, the Tory faction in general was beginning to have some political clout. Arbuthnot, for example, was appointed physician to the queen. Prior to her accession to the throne, Gregory was appointed tutor to Anne's short-lived son, William Henry, Duke of Gloucester. Although Halley had introduced De Moivre to the Royal Society, he was also a friend of David Gregory.[23] That put him in a difficult position in the Cheyne affair. If Gregory did talk to Halley about keeping De Moivre's response out of *Philosophical Transactions*, as Flamsteed claimed, then Halley was siding with Gregory because, at the time, Gregory was more powerful. The Cheyne affair did not damage the friendship between De Moivre and Halley, as may be seen from later social interactions. In 1708 the two met with another mathematician, John Machin, and celebrated the recent victory of the Duke of Marlborough over French forces at the Battle of Oudenarde.[24]

Cheyne published a rebuttal to De Moivre's *Animadversiones* in 1705 under the title *Rudimentorum Methodi Fluxionum Inversæ Specimina: quæ Responsionem Continent ad Animadversiones Ab. de Moivre in Librum G. Chœynei, M.D. S.R.S.*[25] It was not actually a rebuttal but a well-planned retreat with several barbs for his opponent. It was set out using several military metaphors throughout the text; for example, Guerrini translates the title given at the beginning of Cheyne's reply on the first page of the book as "Response to de Moivre's Skirmishes." De Moivre's friend and biographer, Matthew Maty, reported that although De Moivre's criticism of Cheyne was scathing, the "latter's reply carried even more venom in its tail." After reading Cheyne's rebuttal, De Moivre commented in a letter to Bernoulli:

> Mr. Cheyne's book in response to my comments has finally been published. Imagine this, Sir—a man in continual fits of insanity and rage, who spews out crude and ridiculous insults against me at every line: this is the picture of Cheyne in his response to me. Imagine also all the weakness in reasoning, powerlessness and dishonesty that you may, and you will have a fair idea of his book. It is said that Craig had quite a share in it; I am willing to believe that: it is worthy of Cheyne and Craig, the two leading lights of Scotland.[26]

Judging by this outpouring, there may be some substance to the politicking of the Scots in the Royal Society. As late as October 1706, De Moivre was planning to respond to Cheyne's latest publication,[27] but nothing came of it. The dispute upset De Moivre enough that Halley advised him to work on problems in astronomy as a diversion.[28]

Varignon's interest in the dispute between Cheyne and De Moivre predates his May 2, 1711, letter to Bernoulli in which he blamed the Cheyne and De Moivre dispute on Newton. As early as June of 1705, Varignon was writing to Bernoulli to ask him to find a copy of De Moivre's *Animadversiones* for him. Bernoulli wrote back to say he would get a copy for him the next time he was in Amsterdam.[29] Nothing came of it until July of 1706 when De Moivre sent Bernoulli a number of recently published books: Newton's *Opticks*, Halley's translation of works by Apollonius of Perga, Cheyne's rebuttal, *Rudimentorum Methodi Fluxionum Inversæ*, and De Moivre's *Animadversiones*. The last book was meant for Varignon.[30] The path to Varignon was long and winding and used De Moivre's Huguenot connections. De Moivre sent the books via Paul Vaillant in London to The Hague, where Vaillant, through his son, operated a publishing house. Vaillant's office in The Hague sent De Moivre's package to a bookseller in Amsterdam. Bernoulli received the package in Basel late in February of 1707 and sent on the long-awaited copy of De Moivre's *Animadversiones* to Varignon in Paris. Upon receipt and his appetite whetted, Varignon now wanted a copy of Cheyne's *Rudimentorum Methodi Fluxionum Inversæ*.

A Huguenot refugee like De Moivre, Vaillant had fled to London from Saumur in 1686. He soon established himself as a bookseller importing foreign-language books from the Continent.[31] His shop was in the Strand, near Southampton Street, another short walk for De Moivre from his lodgings. This is the first of a few known situations in which De Moivre relied on his Huguenot connections to help strengthen his connections in the Republic of Letters.

Letters between De Moivre and Bernoulli continued to discuss the Cheyne affair into 1707. De Moivre corrected some of the mistakes he had made in his *Animadversiones*. Bernoulli reported that he had received a long complaining letter from Cheyne and had written back reprimanding him. The tempest eventually blew itself out. Varignon reported to Bernoulli in his May 2, 1711, letter that Cheyne and De Moivre had reconciled and were now friends.[32] Later, Cheyne apologized in print for his written attack on De Moivre.[33]

Not all of the early correspondence between De Moivre and Johann Bernoulli was devoted to Cheyne. There were exchanges of several other mathematical ideas and results. For example, in 1705 De Moivre worked, unsuccessfully, on a problem that Bernoulli had suggested in the *Journal des sçavans* in 1703. The problem is to transform a curve, described algebraically, into another of the same length. In addition to De Moivre, several mathematicians erred in their attempts to find a solution. De Moivre made the error "of constructing but a linear transformation of the referent Cartesian axes which leaves the given curve itself untransmuted."[34] De Moivre had shown his solution to Newton[35] who failed to notice the error. It was Bernoulli who pointed out De Moivre's error to him. Eventually, Bernoulli and Leibniz were successful in finding a solution.[36]

This exchange was not entirely devoid of Cheyne. Among the other mathematicians who erred was John Craig, whom De Moivre connected in his own mind with Cheyne's *Rudimentorum Methodi Fluxionum Inversæ*. De Moivre could not resist taking a shot at Craig, and consequently at Cheyne. After celebrating the victory of the Battle of Oudenarde with Halley and Machin, the three settled down to discuss new work that had recently appeared in *Philosophical Transactions*. Halley brought up Craig's published solution to Bernoulli's problem. When Halley described Craig's method of solution from memory, De Moivre claimed it was incorrect since Craig's method of solution was similar to his own unsuccessful attempt. After the three of them consulted a copy of *Philosophical Transactions*, the result was indeed incorrect as De Moivre had predicted. De Moivre quipped to Halley, "Craig has not failed; he will always remain himself, and worthy of having an admirer like Cheyne."[37]

Another exchange between De Moivre and Bernoulli was about centripetal forces. De Moivre wrote to Bernoulli on July 27, 1705, about a discovery he had made that he thought was a new generalization of a result in Newton's *Principia*.[38] The diagram shown to illustrate the result is similar to what De Moivre had drawn in his letter to Bernoulli. Suppose that a planet travelling in an elliptical orbit is at the point P in the orbit as shown. A tangent line to the ellipse is drawn at P. Then the lengths of two more lines and the radius of curvature of the ellipse determine the centripetal force of the planet. The first line, PF, is drawn from the planet's position to the focal point (F) of the ellipse. The second line, AF, is perpendicular to the tangent line. It passes through the focal point and crosses the tangent line at point A.

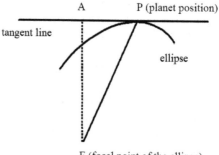

A rendition of De Moivre's diagram for centripetal forces.

De Moivre claimed that the centripetal force is proportional to a^3r/p, where p is the length of the line PF, a is the length of the line AF, and r is the radius of curvature of the ellipse. De Moivre told Bernoulli that he had shown his result to Newton hoping to impress him. However, Newton pulled out an equivalent theorem from among his manuscripts that he was using to prepare the second edition of the *Principia*, which was not published until 1713. Bernoulli wrote back to De Moivre on February 16, 1706, with his own proof while pointing out to De Moivre that the centripetal force was actually proportional to $p/(a^3r)$, the reciprocal of what De Moivre had written.[39]

In a letter to Jacob Hermann dated October 7, 1710, Bernoulli wrote at length about various results in centripetal forces.[40] As part of the letter, Bernoulli informed Hermann of De Moivre's result and said that he had provided De Moivre with a proof of it, without acknowledging to Hermann De Moivre's original discovery.[41] The next month Bernoulli submitted the correspondence with Hermann to the Académie royale des sciences and it was read before the academy in December of that year.

Despite the mathematical uproar with Cheyne and the accompanying upset that it caused him, De Moivre continued to enjoy the company of his Huguenot friends at the Rainbow Coffeehouse next to St. Martin-in-the-Fields Church. According to some 1706 correspondence, these included the three Pierres or Peters: Davall, Coste, and Des Maizeaux. As a sign of the times, Davall complained about their other Huguenot haunt, Pons Coffeehouse up the street. England was in the middle of the War of the Spanish Succession and Pons was now filled with Huguenot officers serving in the English army. The officers often spilled out of the coffeehouse and filled the street causing difficulty for pedestrian traffic.[42]

With visitors and his pupils we have seen De Moivre at his best: witty and sociable. In addition, he could be very helpful to others. For example, in 1706 John Perks, an obscure English mathematician, invented a machine that could be used to find areas under hyperbolas. The idea did not arise in a vacuum; Huygens and Leibniz had worked on other drawing machines that could be used to solve mathematical problems. Through De Moivre's help, the paper describing Perk's machine was published in *Philosophical Transactions*.[43]

Some of De Moivre's helpfulness may have arisen from self-interest or self-promotion. For example, Johann Bernoulli's November 15, 1704, letter to De Moivre contains a reference to Peter de Magneville. It turned out that Magneville was in Groningen at the time and had been there for about three weeks. Bernoulli and Magneville were old friends, and, as Bernoulli found out, so were De Moivre and Magneville. Bernoulli wanted some phosphorus, which had been isolated as an element only about thirty-five years before. Knowledge of the method for extracting it from urine was not widespread, but existed in England through the work of Robert Boyle. Magneville was to bring Bernoulli some phosphorus and Bernoulli asked De Moivre if he would help Magneville obtain it for him. Subsequently, De Moivre obtained the phosphorus and it was sent to Bernoulli by March of 1705. Magneville shows up again in later correspondence. By chance, Bernoulli and

Magneville met in Frankfurt in early 1706 as Bernoulli was making his way to Basel to take up the chair of mathematics; Magneville was on his way to Geneva. A few months later Magneville was visiting De Moivre in London while on his way to Ireland. Magneville reported to De Moivre that the phosphorus had been delivered successfully to Bernoulli.[44]

In other situations, De Moivre could be helpful and sarcastic at the same time. In the 1740s De Moivre and his former student George Lewis Scott helped the composer Johann Christoph Pepusch with the mathematics of ancient Greek music. De Moivre presented Pepusch's paper on the subject to the Royal Society and had the paper published in *Philosophical Transactions*.[45] On the other hand, according to Scott, De Moivre "used to call him [Pepusch] a stupid German dog, who could neither count four, nor understand anyone that did."[46]

The correspondence with Bernoulli, as well as the Flamsteed-Sharp correspondence, perhaps shows another side to De Moivre's character—his combativeness and a touch of cantankerousness. To be sure, he publicly and forcefully pointed out the mathematical flaws in Cheyne's book. His cantankerous side comes out in snippets throughout his life. In an undated letter to Edward Montagu, De Moivre wrote regarding a difficult problem in probability: "I answered Mr. Stevens [another De Moivre student] that I was almost certain the Problem came originally from Robins [Benjamin Robins, a talented and largely self-taught mathematician] and for that reason I would solve it, not to instruct him, but to prove he is a Fool."[47]

Once he had established himself within the Royal Society, and with word spreading of his mathematical interactions with Halley (two early interactions are mentioned in *Philosophical Transactions*), De Moivre soon became the go-to man for mathematical advice and problems that needed solving by other members of the Society. Here are two early examples. While he was listening to De Moivre's complaints about Cheyne's as yet unpublished manuscript in 1702, Flamsteed suggested a geometrical problem for De Moivre to work on. Three months later, De Moivre was still working on it but had not found a solution.[48] A year or two after the Cheyne affair, De Moivre was writing a note explaining some mathematics related to the velocity of comets.[49] The note, dated August 25, 1705, was written on the back of a copy of Halley's 1705 publication on comets. (De Moivre must have followed Edmond Halley's advice to get more interested in astronomy.) The recipient of the note was probably Thomas Sprat, a fellow of the Royal Society and Bishop of Rochester.[50]

Even the Royal Society came to De Moivre in an official capacity to tap into his mathematical expertise. In early 1706, the Reverend Mr. John Shuttleworth wrote to the Society about perspective and an arithmetical method called the Rule of Alligation.[51] Shuttleworth's letter was read before a meeting of the Society; De Moivre was asked to consider Shuttleworth's work and to report back.[52] Much later in the year, in November, Shuttleworth submitted a paper to the Society on perspective. Again De Moivre was asked to look at it.[53] The paper was a critique of a treatise on perspective by the French priest and mathematician Bernard Lamy.[54] In his paper, Shuttleworth claimed that Lamy had not taken into account the position of the person's eyes, especially when viewing an object from an angle. De Moivre read the paper and wrote to Shuttleworth refuting his claim. Not happy with what he received, Shuttleworth wrote to the Secretary of the Royal Society,

> I have sent you Mr. De Moivre's letter. I think he hath not used me candidly in spending so many words upon my letter and saying so little to my treatise. It is, but little encouragement for me to endeavor to perfect the Art of Perspective which L'Amy (tho' a very ingenious author) had not done.[55]

Shuttleworth's paper never saw the light of day in any Royal Society publication. The author published it himself in 1709.[56]

We have come to the year 1707. Abraham De Moivre has turned forty. His mother has died.[57] The eldest child of his brother, a nephew, was born.[58] In the twenty years he had been in England, he has become well respected and established as a teacher and researcher of mathematics with contacts among British mathematicians, the British Whig political elite, and the Republic of Letters. It is time for some kind of mid-life crisis.

❧ 5 ❧

The Breakthrough:
De Mensura Sortis

Shortly after the Cheyne affair, De Moivre became frustrated with England and wanted to leave. The major reasons are fairly obvious, although there is no historical record that ever spells them out. He was an excellent mathematician whom several members of the Royal Society relied upon for mathematical help. He had no permanent job and worked from patron to patron as a tutor to their children, walking across London to each of their houses to give his lessons. He wanted an academic position, or perhaps some regular patronage position, that would give him some stability.

There were two direct roadblocks to achieving his goal. First, to obtain a professorship at Oxford or Cambridge he needed a Master of Arts degree from one of those institutions; he had no degree at all. That problem could be circumvented, but no action was taken at this time. The second roadblock is less obvious and more offensive to someone of De Moivre's talents. As Matthew Maty says,

> he was a foreigner, and frankly, he lacked the kind of savvy needed to win the favour of those who could have ensured that his origins be forgotten and his talent recompensed.[1]

Even though he became a naturalized citizen in 1706, he was still officially considered a foreigner by, for example, the Royal Society. Until 1712 or shortly thereafter, De Moivre appears in lists of fellows among the "Persons of other Nations."[2]

De Moivre had several highly placed Whig connections, but those connections never translated into a patronage position. He also had several scientifically powerful friends such as Newton and Halley, but they did not control the levers of patronage and could only make recommendations to those that did. As a first generation Huguenot refugee, patronage positions were not on the horizon. Although discrimination against talented Huguenots was not confined to De Moivre,[3] there were some advances. Some of De Moivre's Huguenot friends or connections did obtain some second-rung patronage positions, but they were all second generation.

There was also an indirect roadblock. In England there was a general lack of monetary support for science compared to other countries. After he met De Moivre in the late 1730s, the French art critic Abbé Jean-Bernard Le Blanc succinctly summed up the situation, England in general, and De Moivre in particular. First, Le Blanc compared the monetary support for science in England to that of France where the Académie royale des sciences had established a series of prizes. He quipped to his correspondent, the dramatist Pierre-Claude Nivelle de La Chaussée:

> Will you not grant me, sir, that if the English love the sciences better than we, it is strange (I should not say it, if truth did not authorize me) that the only prizes founded here are for horse-racing.[4]

Then Le Blanc finished his criticism of the English support of science by again comparing the English situation, De Moivre in particular, to that of France where members of the Académie were given good pensions. He went on to say to La Chaussée:

> Several Frenchmen will tell you, that at London Farinelli [a famous castrato singer] has gained immense sums in one winter; and they will tell you the truth. Yet all this liberality of the English is but the effect of their ostentation: it is not even a proof of their taste for Italian music. At least, while they pay such high prices to those who excel in an art, that ought to appear frivolous to them; it is surprising that a gentleman, who has rendered himself so valuable to science which they honour most, that Mr. De Moivre one of the greatest mathematicians in Europe, who has lived fifty years in England, has not the least reward made to him; he, I say, who, had he remained in France, would enjoy an annual pension of a thousand crowns at least in the academy of sciences.

The roadblocks combined, both direct and indirect, left De Moivre to fend for himself.

De Moivre's opportunity to obtain an academic position outside England came when Johann Bernoulli decided to leave Groningen in 1705, thus opening the position of professor of mathematics there. De Moivre wrote to Bernoulli, now in Basel, in December of 1707 asking for his help in obtaining the position at Groningen or at another Dutch university situated in the town of Franeker. De Moivre wrote again to Bernoulli in 1708 asking if Bernoulli had contacted Johannes Braunius, Bernoulli's friend and professor of theology at Groningen, about De Moivre's case.[5] Bernoulli wrote to Leibniz about the issue in 1709 but nothing came of any efforts that they made. De Moivre was still looking for an academic position on the Continent as late as 1710. In April of that year, Bernoulli wrote again to Leibniz asking for his advice on positions that might be available to De Moivre.[6] Nothing came of De Moivre's effort, and he remained in England for the rest of his life.

While he was trying to find an academic position, De Moivre continued to work on mathematical problems and to get his work into print in *Philosophical Transactions*. In 1707 he was able to find an algebraic expression for one of the roots of a polynomial equation that has a very specific (only odd powers) and special form (the coefficients are a specified function of the degree of the polynomial).[7] He gave no motivation for why this equation was of any interest to him, other than it had a nice solution. In typical De Moivre fashion, he stated only what the roots were and gave no information about how he obtained the roots.

He revealed some of his methodology in a 1722 paper in *Philosophical Transactions*.[8] As described in that paper, the method involves a trigonometric argument that results in two equations: $1 - 2z^n\cos(n\theta) + z^{2n} = 0$ and $1 - 2z \cdot \cos(\theta) + z^2 = 0$ for some angle θ, where n is the degree of the polynomial in the original equation of interest. The solution to the second equation, a quadratic equation, yields two solutions: $z = \cos(\theta) \pm i \cdot \sin(\theta)$, where i is the imaginary number $\sqrt{-1}$. This is De Moivre's first step into the mathematical area of complex analysis. Substituting these results from the second equation into the first equation yields a single equation in the angles θ and $n\theta$. The unstated methods used in the 1707 paper, but revealed in 1722, form the foundation for major mathematical results he obtained in *Doctrine of Chances* in 1718 and later still in his *Miscellanea Analytica* in 1730. Ivo Schneider has provided a detailed mathematical analysis of all these results.[9]

Other work from this time did not find its way into print. In a 1708 letter to Johann Bernoulli, De Moivre obtained a version of a Taylor series expansion, but did not publish his results. Ten years later the expansion did find its way into *Doctrine of Chances* to sum a finite sequence.[10] The 1708 work was originally motivated by Newton's work in finite differences. Initially, De Moivre used the result to evaluate finite sequences in which some higher order differences on values of the sequence are zero. This could be applied, for example, to finding the sums of cubes of the natural numbers since the fourth difference of the sequence would be zero. His next application of his series expansion was to solve Newton's problem of finding the curve that passes through a given set of points. Finally, he was able to use the method to find the quadrature of a curve that Bernoulli had considered.[11]

De Moivre could have remained in this state indefinitely—producing good, but what might be called second-tier, research when compared to Newton and Bernoulli. But then along came Francis Robartes, who was to change De Moivre's scientific career forever. Robartes was the younger son of an aristocrat, John Robartes, 1st Earl of Radnor. Eventually Francis's son, but not Francis as has often been mistakenly written, inherited the family title and fortune. Francis Robartes was very much in evidence on the political scene. With only a few gaps, he was a Member of Parliament for several constituencies from 1673 (when he was 23 years of age) until his death in 1718. He also held several political patronage positions at various times in his career, including the highly lucrative one of Teller of the Exchequer. At the same time he was active in the Royal Society, serving as vice president for several years during Newton's presidency. As a gentleman virtuoso mathematician, he took on some mathematically related duties in the Royal Society. In 1704 he was chosen as one of the six referees supervising a royal grant for the publication of John Flamsteed's widely subscribed star catalogue *Historia Coelestis Britannica* that was eventually published in 1712. Also in 1712, he sat on the Royal Society committee, as did De Moivre, to determine Newton's priority in the discovery of the calculus.[12] For Newton, the committee's report, which sided with Newton, was the high point in the priority dispute over the calculus.[13]

One of Robartes's mathematical interests was probability. Most of his ideas came from a combination of Newton's binomial theorem and Christiaan Huygens's 1657 treatise, *De ratiociniis in ludo aleae*, which De Moivre had also read when he was a student at Saumur. Huygens had solved some classical problems in probability: the problem of the division of stakes and various problems related to throws of dice. In the division of stakes problem, the initial situation is that two players agree to play a series of games for a pot of money until one of them wins a majority of the games and hence the pot. The wrinkle is what to do, or how to divide the pot, when the series is terminated before the winner is determined. The dicing problems solved by Huygens were of various types:

1. Find the probabilities of the various sums of the faces that show in the throw of three dice.

2. Find the number of throws required to see a six appear at least once with probability 1/2 in the throw of a single die, or two sixes at least once in the throw of two dice.

3. Two players bet on seeing two different outcomes for throws of the dice. In a series of throws of the dice, find the probability that one player will see his outcome appear before the other player sees his.

Huygens's treatise also contains a challenge problem that came to be known as the gambler's ruin problem. Related to the gambler's ruin is another called the problem of the duration of play. The general description of the two problems is that a series of games is played between players of different skills (or different probabilities of winning a game) who have different amounts of capital at the beginning of the series. The winner of any game in the series is given one unit from the loser's capital. The series of games ends when the capital of one of the players has been exhausted, or, in other words, the player has been ruined. The object of the gambler's ruin problem is to find the probability that one player ruins the other, either eventually or within a specified finite number of trials. The object of the duration of play problem is to find the probability that play ends within a given number of games, which is the same as finding the probability that either of the two players has been ruined within that time. Anders Hald has a modern discussion of the various eighteenth-century solutions to these problems.[14]

In 1692, Robartes wrote a manuscript on two probability problems that he presented to the Royal Society but never published.[15] One of the problems is a variation on problem numbered (3) in the list of Huygens's dicing problems. It is also a simplification of a problem posed by Johann Bernoulli in 1685 in *Journal des sçavans* and later solved by both him and Leibniz in 1690 in separate articles in the journal *Acta Eruditorum*.[16] The next year, in 1693, Robartes, thinking he had detected a probabilistic paradox, presented a paper on it to the Royal Society. This one was published,[17] and has been either ridiculed or ignored since the nineteenth century when Isaac Todhunter severely criticized the paper in his *History of the Theory of Probability*.[18]

Todhunter's criticism, and subsequent ones based on it, has been unfair to Robartes. In his tract on probability, Huygens dealt with probability through expected values. An expected value is the weighted average of the possible values of the outcomes, with the weights given by the probabilities of each of the outcomes. Huygens started with the very simple proposition that if he is offered an amount a or an amount b, where either can be obtained with equal facility, then his expectation, or what the offer is worth to him, is $(a + b)/2$. From that initial proposition, Huygens built a mathematical structure that allowed him to solve all the outstanding probability problems of his day and more. For some of the situations that he considered, Huygens gave the odds of winning for each of two players in a game. He calculated the odds as the ratio of the expectations for each player rather than the ratio of the chances to win. For the situations that he considered, Huygens's calculations were correct. Robartes showed that if it is assumed generally that the odds of winning are the ratio of the expected gains that could be obtained by each player then a paradox could be constructed.

After this early and rather elementary work, there is an apparent hiatus in Robartes's visible interest in probability for about fifteen years or so. Then, in 1710 Robartes helped John Harris with his article entitled "Play" that appears in Harris's scientific dictionary *Lexicon Technicum*. Robartes devised an algorithm that Harris used to extract the appropriate terms in a binomial expansion in order to solve the problem of the division of stakes.[19]

At some point over the years 1708 to 1710, Robartes received a copy of a French book on probability bearing the title *Essay d'analyse sur les jeux de hazard*.[20] It was written by Pierre Rémond de Montmort, whose chateau is only about 70 kilometers from De Moivre's birthplace, Vitry-le-François. Shortly after its publication, Montmort sent several copies of the book to English mathematicians, including one each to Isaac Newton and William Jones. Robartes may have received his copy directly from Montmort or he may have obtained it via another route.[21]

At first glance, and as noted by an anonymous reviewer of the book in 1709 in the *Journal des sçavans*,[22] *Essay d'analyse* appears to be written with gamesters in mind. The book is written in three parts. Part I is devoted to a mathematical analysis of card games. A discussion of the combinatorial mathematics needed to carry out this analysis appears at the end of this section; these mathematical methods are also used throughout the rest of the book. Part II is a mathematical analysis of dice games, including board games using dice. In addition to actual dice games, Montmort follows up on Huygens's result for the sum of the faces that show on three dice by giving a table showing the number of chances for each of the sums in the throws of two through nine dice. The third and final part deals with Huygens's *De ratiociniis*. Huygens had set out five challenge problems in his book. Montmort solves them all and then sets four more of his own challenge problems, all of which deal with card games. The methods that Montmort uses in *Essay d'analyse* go well beyond what had been done during the seventeenth century.

After he obtained it, either directly or indirectly from Montmort, Robartes showed his copy of *Essay d'analyse* to Abraham De Moivre. He also gave De Moivre three challenge problems of his own devising to work on. According to De Moivre,[23] Robartes "was pleased to propose to me some Problems of much greater difficulty than any he had found in that Book," the *Essay d'analyse*.

The first two of Robartes's problems were inspired by bowling—lawn bowling, not the more common five- and ten-pin games played in bowling alleys today. Lawn bowling in the eighteenth century had both a rough and a genteel side to it. The rougher side is described by Charles Cotton in *The Compleat Gamester*, first published in 1674 but reprinted in 1709,[24] around the time that Robartes was posing his bowling problems to De Moivre.

> Bowling is a game or recreation, which if moderately used is very healthy for the body, and would be much more commendable than it is were it not for those swarms of rooks which so pester bowling-greens, bares and bowling-alleys where any such places are to be found, some making so small a spot of ground yield them more annually than fifty acres of land shall do elsewhere about the City, and this done cunning, betting, crafty matching and basely playing booty.

The genteel side of lawn bowling, more likely what Robartes had in mind, is shown in a 1738 painting of the gardens of Hartwell House. This is a country estate near Aylesbury, Buckinghamshire, about 75 kilometers from London, owned at the time by Sir Thomas Lee, baronet and Whig politician.

The bowling green and octagon pond at Hartwell House by Balthasar Nebot (from the Buckinghamshire County Museum collections).

Robartes's two bowling problems (Problems 16 and 17 in *De Mensura Sortis*) are a variation on the division of stakes problem. Two players of equal ability are bowling. In the first problem, one player (call him A) needs one point and the other (call him B) needs two points to win. In the second problem, B needs three points to win. In both problems, the object is to find the relative chances that each player has of winning. The twist is in the way that bowling is played. The game consists of several *ends* in which the players, each with an equal number of balls, roll their balls in turn across the bowling green. At each end of play, a jack is thrown down the green and the players bowl toward the jack. The player whose ball is closest to the jack takes the end. That player counts one point for each ball that is closer to the jack than the opponent's best ball. Then the jack is thrown in the opposite direction to begin another end. Ends continue and the game goes up and down the green until a specified number of points have been reached.

How Robartes's first problem differs from the usual division of stakes problem is that in the end being played, A can score one or more points and finish the game, or B can score two or more points and finish the game. If B scores one point only, the game continues another end. De Moivre's method of solution is similar to that used by Huygens to solve the standard division of stakes problem. It also uses Huygens's approach to probability through expectations. De Moivre focuses on player B, the one who needs two points to win. He defines probabilities for two events: B has at least two balls closest to the jack, and B has at least one ball closest to the jack. Having exactly one ball closest to the jack is merely the difference in the probabilities of these two events. For the current end being played, B can win the stake, set at a value one, by having at least two of his balls closest to the jack, or he can even the match by having exactly one ball closest to the jack. In the latter case, from Huygens, the value of the stake is 1/2 since the two sides are even with the game unfinished. De Moivre proceeds to find this player's expected return based on the two defined, but not calculated, probabilities. Since it is based on a stake of value one, it is also the probability that B wins. To obtain the complete solution, De Moivre needed expressions for the probabilities of the two events that he has defined. This he does using permutation arguments. He looks at the restricted arrangements in each of the two cases when one or two of B's balls are closest to the jack with the remaining balls unrestricted, and compares these results to the total number of arrangements of all the balls in play. Using the solution to the initial bowling problem, De Moivre uses a recursive argument in the spirit of Huygens, as well as permutation arguments to solve Robartes's second bowling problem.

Robartes worked on the bowling problem himself and had come up with a laborious solution involving several cases. Once De Moivre had solved the first problem, within a day of Robartes posing it, Robartes gave De Moivre the other two problems to work on (the second bowling problem and a dicing problem), while at the same time encouraging him to write on probability. The encouragement proved fruitful. De Moivre finished his manuscript on probability during a holiday that he spent at a country house, possibly Robartes's.[25] On June 11, 1711, De Moivre submitted his manuscript to the Royal Society. The Society's Journal Book quietly marked the beginning of a new era for probability in England with the note, "Mr. De Moivre presented a Treatise Intituled, de Probabilitate Eventum in Ludo Alea, This Treatise was Ordered to be printed in the Transactions."[26]

The treatise, with the title *De Mensura Sortis* or "Of the measurement of lots," comprises an entire issue (Number 329) of *Philosophical Transactions*. At 52 journal pages, it is more than three times longer than anything else De Moivre had written to that date.

Contrary to the reviewer in the *Journal des sçavans*, another reviewer of Montmort's 1708 *Essay d'analyse*, this time in the *Journal de Trévoux*, stated that the book would give more pleasure to mathematicians than it would be of use to gamblers.[27] It is an interesting contemporary insight in that Montmort carried out mathematical analyses of several games of chance that were popular in his day. After it was published, the *Journal de Trévoux* also reviewed *De Mensura Sortis* and came to the same conclusion, this time saying De Moivre's book would be applauded in the halls of the academy but would be of little use in the halls of gaming.[28] The reviewer also made the point that typically gamblers would not be able to understand the underlying mathematical principles, to follow subtle proofs of results, or to carry out difficult calculations. These observations were mixed in with some moralistic comments on gambling (the *Journal de Trévoux* was a Jesuit periodical that was devoted to the defense of orthodox Roman Catholic doctrines and beliefs) and a précis of the mathematical contents of the paper. Whatever his moral stance, the reviewer was correct in his assessment—*Essay d'analyse* and, to a greater extent, *De Mensura Sortis* are works of mathematics that are only marginally related to the gambling culture of the day.

Although the publication date is given as 1711, *De Mensura Sortis* was not in print until 1712.[29] Shortly after its publication, De Moivre sent copies of the issue to several people in England, including Edmond Halley,[30] Isaac Newton,[31] and De Moivre's fellow chess player at Slaughter's Coffeehouse, the Earl of Sunderland.[32] De Moivre's friend, Pierre Des Maizeaux handled several copies that were bound for the Continent.[33] Using his connections in the Republic of Letters, Des Maizeaux sent copies of *De Mensura Sortis* to Abbé Jean-Paul Bignon, at that time the French minister of state with responsibility for the Académie royale des sciences. Bignon wrote to Des Maizeaux on September 24, 1712, saying that the copies he received had been distributed. He also enclosed a letter from Montmort to De Moivre thanking him for his treatise; the letter has not survived.[34] Whatever he thought personally about De Moivre's treatise, Montmort was adhering to the code of civility in the Republic of Letters by sending the letter of thanks. Other people on the Continent receiving copies were Nicolaus Bernoulli,[35] Johann Bernoulli, and Pierre Varignon.[36] Johann Bernoulli received his copy via William Burnet, a younger son of Gilbert Burnet, Bishop of Salisbury; Bernoulli had asked Burnet to obtain a copy for him.[37]

Montmort was probably offended by what De Moivre had written in *De Mensura Sortis* in his dedicatory preface to Robartes:

Huygens was the first that I know who presented rules for the solution of this sort of problems, which a French author has very recently well illustrated with various examples; but these distinguished gentlemen do not seem to have employed that simplicity and generality which the nature of the matter demands; moreover, while they take up many unknown quantities, to represent the various conditions of gamesters, they make their calculation too complex; and while they suppose that the skill of gamesters is always equal, they confine this doctrine of games within limits too narrow.[38]

Montmort replied at length to this preface in his own preface to his new edition of *Essay d'analyse* that came out in 1713.

In the meantime, about a month after receiving his copy of *De Mensura Sortis*, Montmort wrote to Nicolaus Bernoulli on September 5, 1712. He commented on all the problems in De Moivre's treatise.[39] In nearly every case, Montmort pointed out that either the result was in his 1708 edition of *Essay d'analyse* or that he and Bernoulli had recently obtained the result. The letter to Nicolaus was sent via his uncle Johann Bernoulli. Before sending it on, the elder Bernoulli commented in a letter to Pierre Varignon:

> I have also received Mr Moyvre's treatise *de mensura sortis*; I find his way of solving these kinds of problems a little obscure—I had already solved most of them beforehand in a far easier and more intelligible way. Mr de Montmort has written a letter to my Nephew, care of myself, in which he also mentions this treatise; I have already sent on this letter to my Nephew in England—if you have the opportunity, would you kindly inform Mr de Montmort that I have done so, with my compliments.[40]

Bernoulli made essentially the same comment to William Burnet when he wrote to Burnet to thank him for sending him a copy of *De Mensura Sortis*. There Bernoulli expressed the shortcomings in De Moivre's approach to probability as less simple and less natural than what he had taken.[41]

Montmort's, and to a lesser extent Bernoulli's, criticisms of De Moivre's work are overly severe and unfair. *De Mensura Sortis* was a brilliant beginning for De Moivre. Some of his results were in the 1708 *Essay d'analyse*, but De Moivre either had a different approach or was using them to lay groundwork for later results in his treatise. For the results that Montmort and Nicolaus Bernoulli worked out after 1708, Montmort was trying to claim priority of discovery rather than give credit to De Moivre for independent work. For his part, Johann Bernoulli did not publish any results that he had obtained in probability, so it is difficult to make any comparison of his methods of solution to De Moivre's. Johann Bernoulli was correct about some of De Moivre's solutions being obscure; as we have seen a few times before, De Moivre did not reveal some of his methods of solution.

During September 1712, Montmort must have hidden his anger from the powerful Bignon and therefore maintained decorum within the Republic of Letters. That decorum was broken after he published his letter to Nicolaus Bernoulli in the 1713 edition of *Essay d'analyse*.[42] In addition to telling Bernoulli that there was nothing new in *De Mensura Sortis*, Montmort was critical of nearly every one of De Moivre's results. Montmort used stronger language in a 1716 letter to Brook Taylor saying that De Moivre had plagiarized his work.[43] Writing in his 1719 éloge of Montmort, the perpetual secretary of the Académie royale des sciences, Bernard de Fontenelle, mentioned Montmort's concerns that De Moivre had plagiarized his work. All these accusations were too public for Fontenelle's taste. He strongly promoted the code of civility within the Académie[44] and consequently chided Montmort for his behavior as he lay in his grave.[45] In the éloge, Fontenelle did a double entendre on Montmort's position in society as lord of the manor— Seigneur or Sieur de Montmort—and on his academic position as author of *Essay d'analyse.* After mentioning De Moivre's work and how Montmort was angered by it, Fontenelle compared Montmort to a lord of the manor who demanded loyalty and respect in addition to praise from his tenant De Moivre and then questioned whether Montmort was indeed lord of the manor ("& ne decide nullement s'il étoit en effet le Seigneur").[46] Despite the severity of Montmort's comments and his anger, his letter to Nicolaus Bernoulli can be used profitably as a guideline to put De Moivre's results in *De Mensura Sortis* into some context.

De Moivre begins *De Mensura Sortis* with two basic definitions from which many of his results are derived. The first comes directly from Huygens's *De ratiociniis*. For two players, A and B, contending for a stake of value *a*, A has *p* chances to win and B has *q*. The expected value for each player follows what Huygens obtained: $ap/(p + q)$ for A and $aq/(p + q)$ for B. The second definition may have come from Edmond Halley or Francis Robartes. If an event can happen in *p* ways and fail to happen in *q*, and a second event can happen in *r* ways and fail to happen in *s*, then all the chances for events happening or failing are in the product $(p + q)(r + s)$ or $pr + qr + ps + qs$. For example, *pr* is the number of ways both events can happen and *ps* is the number of ways that the first event happens and the second fails. This is the approach that Halley used in evaluating joint life annuities in his 1693 paper on mortality data from the city of Breslau.[47] It is essentially the same approach used by Robartes in his 1692 manuscript that he read to the Royal Society.[48] De Moivre finishes the introduction by saying that if the first event is repeated *n* times, then the total number of chances in the game is given in the binomial expression $(p + q)^n$. When this expression is expanded, it may be written as a sum containing terms of the form $p^i q^{n-i}$ multiplied by an appropriate coefficient, where *i* represents the number of times the event happens and $n - i$ represents the number of times it fails. For example, in the expansion of $(p + q)^n$, p^n is the number of chances for the event happening all *n* times, and $np^{n-1}q$ is the number of chances for the event happening $n - 1$ times and failing once. The binomial expansion becomes the motif for the paper.

In his *Essay d'analyse*, Montmort had used Huygens's approach of finding expectations to solve probability problems. As can be seen from his initial definitions, De Moivre uses a mixture of Huygens's approach and what is now known as the classical definition of probability to solve his problems. To solve many of his problems, De Moivre enumerated the number of cases or chances favorable to an event and then the number of chances unfavorable to the same event in order to calculate the odds.

The problems in *De Mensura Sortis* follow the structure of a piece of music written in ternary form. Such music has an opening theme, a sharply contrasting middle section, and a return to the theme at the end. The major theme might be called "Meditations on the use of the binomial expansion in probability." De Moivre begins and ends with the binomial motif. At the beginning (nine of the first ten problems) there are some simple variations on the use of the expansion of $(p + q)^n$ and then at the end (the last seven problems) the expansion is used to solve a very complex problem, the problem of the duration of play. In the middle, there are several solutions to a number of challenge problems taken from various sources, including the three from Robartes. De Moivre uses a variety of techniques, other than the binomial expansion, for his solutions in this part.

Taken as a whole, *De Mensura Sortis* shows De Moivre's ability to exploit the use of the binomial expansion in novel ways and his deep insight and ingenuity in finding other methods of solution where the binomial expansion does not apply. The details of the mathematics in *De Mensura Sortis*, some quite intricate, have been described fully elsewhere.[49] Consequently, the focus of the remainder of this chapter, as well as the rest of the book, is to give the flavor of the mathematics and how it fits with the general message that De Moivre tried to convey.

De Moivre considered the use of the binomial expansion for three situations depending on what has been given and what is required. The ingredients to these situations are

1. the probability of the event in question;

2. the number of chances for success p; and

3. the number of trials n.

The three are connected in the following way. The probability of any event, ingredient (1), is given by the appropriate collection of some of the terms in the expansion

$$(p+q)^n = p^n + \frac{n}{1}p^{n-1}q + \frac{n}{1}\frac{n-1}{2}p^{n-2}q^2 + \frac{n}{1}\frac{n-1}{2}\frac{n-2}{3}p^{n-3}q^3 + \cdots,$$

which depends on p and n. Given numerical values for any two of these ingredients, De Moivre came up with a method to find a numerical value for the third. The first five problems in *De Mensura Sortis* introduce the variety of ways of using the binomial expansion in this way.

In the first two problems, p and n are given and the probability of an event is sought. Problem 1 is to find the relative odds of throwing at least two aces in the throw of eight dice. The general solution is to obtain all cases, or $(p+q)^n$, and then subtract the two cases for which no ace (q^n) or one ace (npq^{n-1}) is thrown. The resulting odds are given as $(p+q)^n - npq^{n-1} - q^n$ to $npq^{n-1} + q^n$. Problem 2 is a division of stakes problem. The algorithm used is the same as the one suggested by Robartes for the article on play in *Lexicon Technicum*.

In Problems 3 and 4, the number of trials n is known, the probability of the event is equal to $1/2$ and the odds p/q, which can be expressed as z to 1 or 1 to z as convenient, are sought. These problems are a variation on the division of stakes problem. If player A has a games left to win and B has b games left to win, then the expansion using odds of z to 1 instead of p and q is $(z+1)^{a+b-1}$ since the total number of games left to play will be at most $a + b - 1$. Using the z-notation, the algorithm to separate the winning cases for A from B, given by De Moivre and earlier by Robartes in *Lexicon Technicum*, is to assign all terms in z with a power of a or higher to A and all terms in z with a power of $a - 1$ or lower to B. The solution is obtained by setting the ratio of the winning cases for A to that of B equal to 1 and then solving for z. At that value of z, A and B each have probability $1/2$ of winning the game. This is illustrated here with Problem 4. A and B are to play until one of them has won three games. The twist is that A gives B one game so that for A to win, A must win three games before B wins two. The solution is to expand $(z + 1)^{3 + 2 - 1}$ $= (z + 1)^4$. The expansion can be broken into two parts: the chances for A, $z^4 + 4z^3$; and the chances for B, $6z^2 + 4z + 1$. The solution to the equation $z^4 + 4z^3 = 6z^2 + 4z + 1$ yields the required odds. De Moivre says, "z will be found to be 1.6 very near." Using modern computing, the positive root of the equation is 1.592503317 to nine decimal places.

In Problem 5, interest is shifted to finding n, given a value for the chance for success p and given that the probability of the event is equal to $1/2$. The problem is to find the number of trials required so that the probability of obtaining at least one success is equal to $1/2$. This is in the spirit of problem (2) in the list of Huygens's dicing problems in *De ratiociniis*. De Moivre's solution reduces to solving for n in the equation $(p + q)^n = 2q^n$. The problem and solution were not at all new; it was just an illustration of the use of the binomial theorem to solve probability problems.

Montmort was not at all impressed by De Moivre's first five problems. He wrote to Nicolaus Bernoulli that he had already used the binomial expansion in 1710 and that he had solved the first five problems in one way or another in his 1708 book. Of course, he did not know that Robartes previously had made use of the same expansion a year or two earlier, or that John Arbuthnot had beaten them all to the solution to Problem 5 by fifteen years or more in an unpublished manuscript.[50] It is likely that De Moivre got the idea for the use of the binomial expansion from Robartes and in the paper was demonstrating to him the power of the technique to solve a variety of problems beginning with solutions that were already familiar to Robartes.

Having established the use of the binomial expansion to solve some standard problems, De Moivre goes off in a new direction with it. For the next couple of problems (Problems 6 and 7), De Moivre extends Problem 5 from finding n, the number of trials that will produce at least one success with probability $1/2$, to finding the n that will produce at least two or three or more successes with probability $1/2$. Contrary to Problem 5, there is no longer a simple solution to Problems 6 and 7. As in Problems 3 and 4, De Moivre reduces the terms p and q to one value z, but in this case uses odds of 1 to z. Using this notation, the binomial expression to obtain any required probability is given by

$$\left(1 + \frac{1}{z}\right)^n.$$

If the chances for A are associated with 1 and for B with $1/z$, then the terms in the expansion that express no success for B (at least one for A) is 1, for none or one success for B (at least two for A) is $1 + n/z$, for up to two for B (at least three for A) is $1 + n/z + n(n-1)/(2z^2)$, and so on. The required number of trials to find, for example, the number required to get at least three successes for A is the solution for n in the equation

$$\frac{1+\dfrac{n}{z}+\dfrac{n(n-1)}{2z^2}}{\left(1+\dfrac{1}{z}\right)^n}=\frac{1}{2}.$$

Only in a few special cases is there an explicit solution for n. When $z = 1$, the expansion of $(1 + 1)^n$ is symmetric with an even number of terms when n is an odd number. Consequently, to obtain a probability of $1/2$ for the event to occur, the event "at least once" requires only one trial, "at least twice" requires three trials, "at least three times" requires five trials, and so on. When z is greater than 1, the only situation in which there is an explicit solution is for the event occurring at least once, i.e., the solution to Problem 5. This did not deter De Moivre, a very able and creative mathematician. He looked at the extreme situation in which n and z are both infinite, but the ratio $x = n/z$ is finite. In this case, terms such as $n(n-1)/z^2$ and $n(n-1)(n-2)/z^3$ in the above equation become x^2 and x^3 respectively, and $(1 + 1/z)^n$ becomes an infinite series in x. This is the first expression of what is now known as the Poisson approximation to the binomial. After some further manipulation, the resulting equation can be solved numerically for x. Then a solution for n can be obtained in terms of a numerical value for x times z. For the event of at least three successes, the equation to solve is

$$\ln(2)+\ln\left(1+x+\frac{x^2}{2}\right)=x.$$

Using modern computing, the solution is $x = 2.674060314$ to nine decimal places, while De Moivre obtained 2.675. Following his typical pattern, De Moivre gives no hint about what numerical method he used to obtain his number.[51] De Moivre does provide some numerical examples with dice. In the spirit of De Moivre, suppose that two dice are thrown and the desired outcome is two sixes. How many times should the dice be thrown so that the probability of throwing the outcome at least three times is $1/2$? For this example, $z = 36$ and the solution for n is $(36)(2.674060314)$, or 96.2. The exact probabilities when n is 95, 96, or 97 are 0.493527, 0.500495, and 0.507414, respectively.

This is one of the places where Montmort was grudgingly impressed by De Moivre's work. He commented that the problems are very ingenious, but was not sure if the numerical results are "perfectly just" and wondered if there might be another way of looking at the problem. By "ingenious," Montmort probably meant that the problems were tricky and that he had no clue as to the method of solution. Unknown to Montmort, and to De Moivre at the time, De Moivre made a minor error that had no effect on his numerical solutions. He was able to express the equation in x, given above, as a fluxional equation, or in modern terminology, a differential equation. He thought the solution to the fluxional equation could be expressed in an infinite series expansion to obtain the solution. He was incorrect and later dropped the fluxional equations from all editions of *Doctrine of Chances*.[52]

Generalizing upon the opening theme of the binomial expansion, De Moivre next looked at the problem of the division of stakes for three players. Problem 8 first considers three players A, B, and C who have a, b, and c games, respectively, left to win to gain the stake, and whose chances are p, q, and r, respectively. The solution can be obtained by collecting appropriate terms in the expansion of the multinomial $(p + q + r)^{a+b+c-2}$. De Moivre states the problem in terms of a numerical example rather than the general expression given. There is a further generalization of the problem to any number of players that can be obtained by increasing the number of terms in the multinomial and changing the exponent.

Montmort again claimed priority for the solution to this problem and pointed to the weakness in his and De Moivre's solutions. Both mathematicians found it necessary to consider a number of cases, with the result that the solution can be cumbersome. The cases arise by having sometimes to assign part of the value of the coefficient of $p^i q^j q^k$ to each of A's, B's, and C's chances. For example, in Problem 8, De Moivre considered the specific case when $a = 1$, $b = 2$, and $c = 3$. One term in the expansion of $(p + q + r)^4$ is $6p^2q^2$. Since A must win one game before B wins two, of the six cases the only one favorable to B is winning the first two games, i.e., the permutation $qqpp$ that makes up one of the terms in $6p^2q^2$. For all other permutations, A wins.

A series of problems then follow which are solutions to four of the five challenge problems in Huygens's *De ratiociniis*. They are not covered in the order that they appear in Huygens. The gambler's ruin problem (Problem 9) is solved first since it is related to the theme of the binomial expansion. For players A and B, with initial capital a and b and chances p and q of winning a game, De Moivre finds that the ratio of A's ruin probability to B's is

$$\frac{p^b(p^a - q^a)}{q^a(p^b - q^b)}.$$

De Moivre had a very elegant solution to this problem that does not use the binomial expansion; Anders Hald calls the proof "an ingenious trick."[53] De Moivre then slightly tweaked the gambler's ruin problem so that the solution to the tweaked problem can be obtained through a binomial expansion (Problem 10). Assume $a = b$ and that the total capital is $2a$. At each turn, A has p chances to win a unit of capital and B has q chances. The stake is won by the first player to obtain a units. The solution is to expand the binomial $(p + q)^{2a-1}$ and to divide the expanded terms into two parts. The first a terms are A's chances and the second a terms are B's.

Montmort liked De Moivre's solution to the gambler's ruin problem and made no comment on De Moivre's tweaked problem. In order to diminish De Moivre's work, Montmort pointed out that Johann Bernoulli had sent Montmort the same solution to the gambler's ruin problem in a letter dated March 17, 1710. It is unclear if Montmort meant the same method of solution or the same final result. Of the three remaining challenge problems in *De ratiociniis* that De Moivre solved, Montmort commented that he did not understand why De Moivre was working on such easy problems that have already been solved previously (by Montmort himself in 1708).

Although set in the context of games of chance, the problems that De Moivre has solved to this point are unrelated to any actual games played. As noted by the *Journal de Trévoux* reviewer in 1712, *De Mensura Sortis* has the appearance of a work that solves interesting mathematical problems rather than gambling problems or problems of strategy of play. However, set in the middle of *De Mensura Sortis* are two problems that come directly from actual games. One is known as either the problem of the pool or Waldegrave's problem (Problem 15), and the other is the problem inspired by lawn bowling (Problems 16 and 17), suggested by Robartes which we have already seen.

Montmort admitted that the lawn bowling problem is one that he had not seen before. However, he could not contain his criticism. De Moivre had said in his dedicatory remarks to Robartes at the opening of *De Mensura Sortis* that Montmort and Huygens did "not seem to have employed that simplicity and generality which the nature of the matter demands." In response, Montmort mocked De Moivre by throwing this statement back in his face. The bowling problem, he said, could have been stated more generally; the players could have an unequal number of balls. This assumption makes no sense in a regular game of lawn bowling, which again points to the primary interests of Montmort and De Moivre—mathematics, not gambling. Montmort makes this assumption and goes on to give a solution. Then he surmised that the problem would be much more difficult if unequal skills between the players were assumed, but he himself provided no solution at that point. Montmort did obtain the general solution by the time the second edition of *Essay d'analyse* went to print.[54]

The problem of the pool in England comes from a card game for two players called *Piquet*. It was a method to bring a third player into the game. The normal game is played with a 36-card deck, the standard deck with the twos through fives removed. There is a scoring system based on the play of the cards; the first player to reach one hundred points wins the game. Although the card game dates from at least the mid-seventeenth century in England, the variation in play through the pool probably dates from the early eighteenth century.[55] Here is a 1719 description of the pool from a popular gambling manual:

> The Pool is another way of playing *Picquet*, only invented for Society; it is in every Way play'd the same with the other Game; but is a Contrivance to bring in a third. As for example, Three persons are to cut, he who cuts the highest Card, stands out the first Game, for it is held an Advantage to be out first. Then the others are cut for Deal, as is before directed; If they play for Guineas, they are to lay down a Guinea apiece, which makes three Guineas: then he who loses the first Game lays down a Guinea more, and goes out, and he who stood out before, sits down; if the first Gamester beats him also, he sweeps the Board, which is called winning the *Pool*; and the loser must lay another guinea to it. But if he who won the first Game, loses the second, he pays his Guinea, and makes room for the other; thus it goes round sometimes, till the *Pool* amounts to a great Sum. You must observe, the *Pool* is never won, till one Person gets two Games successively. Every Person that loses a Game, lays down a Guinea to the *Pool*. When any person is lurch'd at this Play, he lays down one Guinea to the *Pool*, and pays another to him who lurch'd him.[56]

Since the game was "only invented for Society," or the upper classes, and there is no mention of anyone suggesting this problem to De Moivre, it is possible that De Moivre played the game with some of his upper-class associates.

The 1719 description of the pool illustrates the penchant among mathematicians to simplify, and then to generalize, a practical problem. On the simplification side, a player is lurched when he or she fails to obtain fifty points before the opponent obtains the hundred points required to win.[57] Including that possibility in a mathematical analysis would complicate it substantially; De Moivre makes no mention of lurching. At each round of the pool, one guinea, or at least one monetary unit, is put into the pot or pool. De Moivre generalizes this to an amount s of money. His solution to find the chances of winning for each of the three players requires the use of an infinite series. The solution shows the gap at this time between the gambling public and the mathematicians. The author of the 1719 gambling manual says that "it is held an Advantage to be out first." However, in 1711, and later in *Doctrine of Chances* in 1718, De Moivre shows that the player staying out the first round is at a disadvantage when the fine, or the amount put into the pool on losing a round, is less than 7/6 of a unit. De Moivre may have heard the general talk of the advantage of being out first. In 1718, he explicitly gives the expected loss for the player staying out the first round when the fine is one unit.[58]

Once again, Montmort comments that he had already solved this problem, referring to a letter to Nicolaus Bernoulli dated April 10, 1711. For the case when $s = 1$ and using the essentially same method as De Moivre, Montmort did indeed solve the problem. De Moivre had commented in *De Mensura Sortis* that the solution could be extended to more than three players in the pool. Along with his claim to priority, Montmort retorted that De Moivre's method of solution was impracticable for more than three players.

Montmort's problem of the pool was suggested to him by another Englishman, Charles Waldegrave, son of a baronet and brother of a baron who married an illegitimate daughter of James II of England. Waldegrave was a Jacobite living in France at this time.[59] Even though Montmort himself called it the problem of the pool, it has since come to be known as Waldegrave's problem. Although of English aristocratic background and hence a prime candidate to study with De Moivre, it seems that Waldegrave was unknown to De Moivre. When writing to Brook Taylor in 1717 on his forthcoming *Doctrine of Chances*, De Moivre's only reference to Waldegrave is in the phrase "Monmort and his friend at Paris as well as young Bernoully."[60]

Before returning to the theme of solving probability problems through the binomial expansion, De Moivre dealt with one more problem, now known as the occupancy problem. This is Robartes's third challenge problem to De Moivre. In *De Mensura Sortis*, it is set as a dicing problem. A die with $f + 1$ faces is thrown n times. What is required is to find the probability that a specified subset of the faces, g in number, each shows at least once in the n throws. De Moivre used an inclusion-exclusion algorithm to solve this problem. For simplicity of explanation, suppose the faces that we want in the subset are ace, two, three, and so on. The total number of throws when all faces are considered is $(f + 1)^n$ and the total number of throws with the ace removed is f^n. Consequently, the total number of throws with at least one ace is $(f + 1)^n - f^n$. Now remove the two. The number of throws of at least one ace with the two removed is $f^n - (f - 1)^n$. Since $(f + 1)^n - f^n$ is the number of throws of at least one ace with the two in, then the number of throws of at least one ace and at least one two is $(f + 1)^n - f^n - (f^n - (f - 1)^n)$ or $(f + 1)^n - 2f^n + (f - 1)^n$. This can be built up successively until we reach the desired number of faces, g. De Moivre expresses the general solution in terms of powers with alternating signs $(f + 1)^n - f^n + (f - 1)^n - (f - 2)^n$, continued to the last term, $(f - g + 1)^n$. With some foreshadowing of more binomial expansions yet to come, De Moivre states that the coefficients of each of the terms are obtained from the binomial expansion $(1 + 1)^g$. As a variation to the problem, De Moivre looked to solve for n in the problem, given values for f and g, when the resulting probability is $1/2$. Here De Moivre made a simple approximation in order to come up with a simple expression for n.

Once again, Montmort was not particularly impressed with the solution, claiming that the solution could be obtained from a formula he sent to Johann Bernoulli in 1710 along with a proposition in the 1708 edition of *Essay d'analyse*. Harking back to De Moivre's comment on generality, Montmort chided De Moivre by saying that his own solution is more general in that he had obtained the probability of the chosen faces each showing a certain number of times rather than at least once. Montmort did like De Moivre's twist to the problem to solve for n, the number of throws, but he was not sure whether the approximation was good enough.

The last seven problems (Problems 20 through 26) of *De Mensura Sortis* give the rousing finale to the piece expressed by the binomial motif. The finale is the complete solution to the problem of the duration of play. To recap, in the gambler's ruin problem, A and B with capital a and b play until one of the player's capital is exhausted; the probability of ruin of each of A and B is sought. In the duration of play problem what is sought is the probability that play ends, or that someone is ruined, before a stated number of games.

In the case when $a = b$, De Moivre recognized that the number of games must be at least a and so he calculated the probability that ruin will occur after $a + d$ games. The algorithm to obtain the probability is very simple. First expand $(p + q)^a$ and remove the extreme terms p^a and q^a. Multiply the result by $(p + q)^2$, collect terms, and again remove the two extreme terms. Repeat the process of multiplication by $(p + q)^2$ and removal of the extreme terms $1/2d$ times. If d is an odd number, then replace d with $d - 1$. The required probability is the set of terms obtained from the algorithm divided by $(p + q)^{a + d}$.

De Moivre adjusted his algorithm to take into account the situation in which $a \neq b$. Rather than starting at $(p + q)^a$, he started at $(p + q) \times (p + q)$. At each step he removed the terms for which the exponent of p exceeds the exponent of q by b and for which the exponent of q exceeds the exponent of p by a, and multiplies the result by $p + q$. The number of multiplications is the number of games bet on minus one.

Some of the last seven problems are variations on the use of these two algorithms. For example, once he has given the algorithm to solve the duration of play problem when $a = b$, the next two problems (Problems 21 and 22) are to solve a twist on the problem. In this case, a and d are given, as well as the probability that play will last beyond $a + d$ games. What remains to be found is the ratio of chances, or the odds, $z = p/q$ that will achieve the given probability for the duration of $a + d$ games.

Montmort is complimentary yet condescending in his assessment of De Moivre's solution to the duration of play problem and in his general assessment of De Moivre's work. Adjusting Montmort's mathematical notation to conform to what has been used here, here is what he says to Nicolaus Bernoulli:

I am happy to see that Mr Moivre has solved this problem in its entirety, and that his solution agrees perfectly with our own. However, I have great difficulty in understanding how the learned Geometer arrived at this method for calculating $p + q$ to the power a, removing the extreme terms of this product, and multiplying the remainder by the square of $p + q$, and then repeating this process as many times as there are units in $1/2d$. A solution of this form surprises me, all the more since the Author, who having assumed equal chances for Peter and for Paul, once he assumes the chances are not equal, is obliged to take another route, whereas according to me and you the method is the same for the general and particular cases. Nevertheless, I hold this discovery, and his entire Work, in high regard, and I congratulate myself on having opened the way to it by pioneering this approach. I first found it unusual that he has filled this work with the same things we had discussed in our Letters, but it is only natural, since he based his Work on my own, and wanted to pursue these subjects, that the same ideas came both to him and to us. I would only have wished, and it seems that fairness demands it, that he had frankly acknowledged what I have the right to claim in his Work.[61]

Later when he was writing the second edition of the *Essay d'analyse*, Montmort criticized De Moivre's results on the duration of play problem by saying that the result that he and Bernoulli had obtained was simpler and easier to calculate.[62]

Although we have run through the numbered problems in *De Mensura Sortis* more or less in order and have seen what Montmort thought of the results, there remains one important problem that is not one of De Moivre's problems. Placed between Problems 5 and 6 is a lemma that does not seem to fit. The lemma reads in translation, "To find the number of chances by which a given number of points may be thrown with a given number of dice."[63] Put symbolically, the problem is to find the number of ways of obtaining the sum s of the faces that show in the throw of n dice with f faces. The answer that De Moivre gives, for $r = s - 1$ is

$$\frac{r-1}{1} \times \frac{r-2}{2} \times \frac{r-3}{3} \cdots$$

$$-\frac{r-f-1}{1} \times \frac{r-f-2}{2} \times \frac{r-f-3}{3} \cdots \times \frac{n}{1}$$

$$+\frac{r-2f-1}{1} \times \frac{r-2f-2}{2} \times \frac{r-2f-3}{3} \cdots \times \frac{n}{1} \times \frac{n-1}{2}$$

$$-\frac{r-3f-1}{1} \times \frac{r-3f-2}{2} \times \frac{r-3f-3}{3} \cdots \times \frac{n}{1} \times \frac{n-1}{2} \times \frac{n-2}{3}$$

$$+\cdots.$$

where the "..." means, in De Moivre's words, that "the series ought to be continued until any of the factors are either equal to zero or negative." This is another instance of what Johann Bernoulli complained about: "I find his way of solving these kinds of problems a little obscure." De Moivre gives no method of solution, just the answer. And the method of solution is not obvious. It was probably for this reason that he called it a lemma, making a pun in Latin. A lemma in mathematics is usually interpreted as a proposition that is either subsidiary to a main result or preparatory for it. The classical meaning of lemma is "a subject for consideration or explanation."[64] De Moivre had given the answer to a difficult problem for which he had found a simple solution. He was now challenging others to find the solution.

Without knowing how De Moivre solved the lemma, Montmort dismissed the result. He claimed, rightly, that the statement of the lemma was taken from page 141 of the 1708 edition of *Essay d'analyse*. He did acknowledge in the letter to Nicolaus Bernoulli that the result was "curious," perhaps meaning that the answer was interesting. But then he went on to tell his correspondent that he had already sent his own solution to this problem to Johann Bernoulli in a letter dated November 15, 1710, thus claiming priority for the solution.

De Moivre's initial solution to the lemma is explained in a letter to Brook Taylor in 1718.[65] After receiving his copy of *Doctrine of Chances*, where the lemma also appears without proof, Taylor wrote a letter to De Moivre, which has not survived. In De Moivre's reply, dated September 29, 1718, it is clear that Taylor had asked De Moivre about the lemma. De Moivre briefly outlined how he had discovered the proof. It was a combinatorial solution whose method of proof was different from Montmort's and Bernoulli's. For three dice, De Moivre wrote down the number of ways each of the sums could occur. Beside those numbers he wrote the difference between the numbers and the number of combinations obtained by selecting two objects from r, where, again, $r = s - 1$ and s is the sum on the faces that show. Then he looked for patterns in these differences. Upon obtaining the solution for $n = 3$, which agrees with the expression given in the lemma, he went on to $n = 4$ and related to Taylor that he had obtained the result by induction. The first step in the proof, with added detail but only up to 11 points showing on the dice, is shown in the table below. In the table, the combinatorial symbol

$$\binom{r}{t} = \frac{r!}{t!(r-t!)},$$

where $r! = r(r-1)(r-2) \ldots 3 \cdot 2 \cdot 1$ is the product of the first r natural numbers, stands for the number of ways of choosing t objects from r. Later in the 1720s, De Moivre discovered the method of generating functions, which provides a very simple and elegant solution to the problem.

De Moivre was proud of what he had written in *De Mensura Sortis*.[66] He must have been shocked by Montmort's reaction; he was able to see the letter and make a copy of it after Nicolaus Bernoulli arrived for a visit to London in October 1712.[67] When De Moivre wrote to Johann Bernoulli in December 1712, he was circumspect and polite. He had to be—Montmort was also corresponding with the elder Bernoulli and working with Nicolaus. De Moivre listed Montmort's general complaints about his work and commented that Montmort did not understand what he had written in the preface to *De Mensura Sortis*. He assured Johann Bernoulli that he did not want to offend Montmort and that any criticisms that he made were done in an indirect way.

De Moivre had learned his lesson about politesse in the Republic of Letters. For his part, Montmort was unfair in most of his criticisms of De Moivre's work.

Points	Chances		
3	1		$\binom{2}{2}$
4	3		$\binom{3}{2}$
5	6		$\binom{4}{2}$
6	10		$\binom{5}{2}$
7	15		$\binom{6}{2}$
8	21	$\binom{r}{2} - \text{error}$	$\binom{7}{2}$
9	25	$28 - 3$	$\binom{8}{2} - 3$
10	27	$36 - 9$	$\binom{9}{2} - 3$
11	27	$45 - 18$	$\binom{10}{2} - 3$

De Moivre's calculations for his lemma.

Montmort was trying to protect his own personal turf. In Montmort's mind, his latest work in probability could only be the product of his own talent and not something that could be independently discovered by others.[68]

Although a good and respected mathematician, Montmort can be seen as a bit of a dilettante; mathematics was more a serious hobby to him than a full-time occupation. In that way he might be compared to De Moivre's friend and probable aristocratic patron Francis Robartes, who seriously dabbled in science, but did not achieve success to the extent that Montmort had. Parts of Montmort's life point to his dilettantism. He made three trips to England, the first to escape his father. On the last trip in 1715, he went to observe a solar eclipse. Halley's account of the eclipse seems to imply that Montmort was there for the show, rather than as a participant taking scientific measurements.[69] When Montmort was young, he joined the Church but left it when he fell in love with his neighbor's niece and married her.

Prior to his marriage in 1706, Montmort had inherited money, acquired noble status, and set himself up on an estate at Montmort. His marriage gave him some interesting connections. Montmort's new aunt by marriage was the duchesse d'Angoulême, whose husband was the illegitimate son of Charles IX. The duc and duchesse married when he was 72 and she was 23. Montmort's marriage into this family gave him some connections to the royal court. Unfortunately for Montmort, Louis XIV did not like the aunt; her husband was a Valois rather than Bourbon and, consequently, Louis ignored her.[70] The duchesse lived with Montmort until her death. The *Essay d'analyse* is written around games of chance played at the court of Louis XIV.[71] The frontispiece of the book, shown in Chapter 7, depicts the court at play in the background. The book, undoubtedly printed at Montmort's expense, may have been published, in part, to curry favor at court, which never materialized.

I am left with a feeling that the French nobleman was a little miffed by the upstart bourgeois French refugee of the wrong religion who claimed that his results were more general and simpler to obtain. In time, De Moivre responded by obtaining more and better mathematical results and by throwing some subtle and hidden barbs at Montmort that would follow the letter of the law for the code of civility in the Republic of Letters.

∽ 6 ∾

A Newtonian Intermezzo

When Martin Folkes died in 1754, he left a large library of over five thousand books and a small number of manuscripts, which were put up for sale. Among the manuscripts was one that the sale agent described as, "The formation of a catalogue of curves, contain'd between page 62 and 63 of Sir Isaac Newton's tract of quadratures, wrote by Mr. De Moivre, 4to."[1] It sold for six shillings and sixpence. The catalogue of curves, which is essentially a table of integrals, appears between pages 62 and 63 of Newton's *Analysis per Quantitatum Series, Fluxiones, ac Differentias*, which was published in 1711 and edited by Newton's and De Moivre's friend William Jones.[2] Much of the book is based on transcriptions of some of Newton's letters and scientific papers that had been in the possession of John Collins. Jones obtained Collins's manuscripts in about 1708, fifteen years after the latter's death.[3] The book contains some of Newton's earliest work on the calculus and was connected to the dispute between Newton and Leibniz over the discovery of the calculus. In the preface, Jones lays out a short history of the development of the calculus. Rupert Hall and Laura Tilling comment about the preface:

> Without mentioning Leibniz or the calculus dispute, Jones in eleven short pages presented powerful evidence of Newton's mathematical originality as far back as 1665. For the first time the testimony of Newton's earliest communications to Barrow and Collins…was set before the public, thus anticipating the fuller documentation attempted in the *Commercium Epistolicum*.[4]

The *Commercium Epistolicum* is a collection of letters assembled by Newton to support his case against Leibniz. Compared to John Keill's published attack on

Leibniz, appearing in 1710, that contained accusations of plagiarism,[5] Jones's contribution to the priority dispute was much more restrained and muted.

As tempting as it is to conjecture that Jones asked De Moivre to compose a table of integrals for the book, thus making De Moivre a ghostwriter for some of Newton's mathematical work, the cataloger for the sale of Folkes's book made an error. The table appears in some of Newton's early manuscripts written prior to De Moivre's arrival in England.[6] Instead of originating the table, De Moivre probably copied it from Newton's manuscripts on quadrature that he had seen in 1702;[7] it would have been a handy reference for him in view of his research interests at the time. There are other possibilities. De Moivre could have copied the table from the 1711 book. He could also have seen the table in Newton's *Opticks*, published in 1704, or in the 1706 Latin translation of it, both of which contained Newton's treatise on quadrature.[8]

The controversy over the discovery of the calculus that erupted in 1710 had been brewing for several years; the first brickbat against Leibniz from the Newtonian side was thrown by Fatio de Duillier in 1699 in a pamphlet on the brachistochrone problem.[9] The controversy came to a head in a formal way in 1711 when Leibniz, in his capacity as a fellow of the Royal Society, wrote to Hans Sloane, secretary to the Society, demanding an apology from Keill for accusing him of plagiarizing Newton's work. No apology from Keill was forthcoming. Rather, Keill counterattacked by laying out his case in a letter to the Royal Society, which the Society decided to send on to Leibniz. Leibniz wrote back to the Society late in 1711 with a dignified response, giving credit to Newton for his discoveries and at the same time claiming equal credit for his own work. Certainly Keill was excessive and unfair in his attacks on Leibniz who, in truth, had independently discovered the calculus. After 1711, Leibniz reached the limits of his tolerance and with his supporters began a game of tit for tat with Newton and his group. As the whole affair ground on, no one escaped without some mud splatter sticking to them.[10]

Up to the end of 1711, De Moivre remained in the background to the point of invisibility in the dispute. A few years prior to Keill's outburst, De Moivre had praised both Leibniz and Johann Bernoulli for their mathematical abilities. In a letter to Johann Bernoulli, De Moivre began by commenting on John Craig's work that appears in *Philosophical Transactions* for 1708. He mentioned Craig's criticism of Bernoulli's and Leibniz's solutions to a problem that Craig thought he had solved, and then stated,

> The whole world knows, apart from an ignoramus like him [Craig], that Mr. Leibnitz and yourself have pushed the field of mathematics infinitely higher than Mr. Huygens, and that if that excellent mathematician were still alive, he would not hesitate to do you that justice.[11]

Perhaps De Moivre was still feeling the sting of the Scottish faction in the Royal Society, or perhaps it was his cantankerousness coming to the fore again.

De Moivre began to lose his invisibility in the dispute when Newton appointed him to the Royal Society committee to determine Newton's claim to priority in the discovery of the calculus. In reality, De Moivre probably had very little to do with any deliberations of the committee or with writing the report. Newton wrote the draft of the report.[12] Furthermore, the committee was originally established on March 6, 1712. De Moivre was not added to the committee until April 17 and the report, presented on April 24, 1712, to the Royal Society, one week after De Moivre's appointment, was unsigned by any of the committee members.[13] Subsequently, a comprehensive case for Newton from the British point of view was laid out in print in the *Commercium Epistolicum*. It was distributed free to selected individuals over January and February of 1713; De Moivre sent some copies to France.[14] Since the names of the committee members did not appear in the publication, De Moivre could still aspire to anonymity and relative neutrality in the dispute. What is also interesting, from the point of view of De Moivre's scientific connections, is that of the other ten committee members, half of them were good friends of De Moivre or were soon to be: Edmond Halley, William Jones, John Machin, Francis Robartes, and Brook Taylor.

Between the submission of the report to the Royal Society (April 1712) and the publication of the *Commercium Epistolicum* (February 1713), Johann Bernoulli's nephew Nicolaus visited London. He arrived in September 1712 on one of the legs of a tour that took him to England, Holland, and France. De Moivre met him early on in London, and introduced him to both Newton and Halley. De Moivre brought Bernoulli three times to meet with Newton, and twice they were invited to dine with him. Uncle Johann was very pleased with the hospitality his nephew had received from Newton, Halley and De Moivre.[15]

There were no meetings of the Royal Society until mid-October, so Nicolaus Bernoulli was unable to attend a meeting until that time. Also, meetings were closed to non-fellows; Bernoulli could only attend as another fellow's guest[16] or with permission of the meeting's chair, often the president. When meetings recommenced, De Moivre was tied up with some unstated business; it was Halley who introduced Nicolaus Bernoulli to the Royal Society.[17]

From one viewpoint, Nicolaus Bernoulli's trip was a great success. De Moivre wanted both uncle and nephew proposed for fellowship in the Royal Society. Newton felt that it would give greater honor to Johann Bernoulli to be proposed first. This was done on October 23, 1712; he was elected fellow on December 1. Nicolaus Bernoulli's fellowship came about a year and a half later. On February 18, 1713, Johann Bernoulli wrote to De Moivre thanking him for his efforts in obtaining the fellowship.[18]

From another viewpoint, the trip was an indication of storm clouds on the horizon. Nicolaus Bernoulli informed De Moivre of a problem that he had found in one of the results in Newton's *Principia*, in particular Book II, Proposition 10 dealing with resisted motion. The theory was not quite correct, and, consequently, Newton's final numerical result was off by a factor of 3 to 2. It was actually the uncle who had discovered the problem. After De Moivre informed Newton of the problem, Newton worked for two days fixing it and then sent his correction to Roger Cotes, who was in the process of preparing the second edition of the *Principia* for publication.[19] What seemed a relatively minor point at the time became a sore spot in the future, with connections to the priority dispute.

Although Johann Bernoulli supported Leibniz in the priority dispute, he tried to downplay his support in public and to maintain a good relationship with mathematicians in England. One of the things that worried him was that by March 1714 he had not heard from De Moivre for over a year. De Moivre's last letter to him was written in December 1712, when De Moivre gave news of Bernoulli's election to the fellowship. Bernoulli had written to De Moivre in February 1713 and had received no reply. In De Moivre's December letter, he had promised to send Bernoulli a copy of the *Commercium Epistolicum*, and in an earlier letter of October 1712, he had also promised to send the latest edition of Newton's *Principia Mathematica* when it came off the press.[20] Nothing had been received. Bernoulli went to great lengths to find out what had happened. He checked with Varignon about whether De Moivre had received his letter of February 1713. Varignon replied that the letter had been sent through Paris to London to John Arnold's brother, whose valet had delivered it to De Moivre. John Arnold was a former student of Bernoulli and, contrary to almost all British mathematicians, took Leibniz's side in the calculus dispute. On March 8, 1714, Bernoulli wrote to John Arnold, telling him what he had learned from Varignon and asking Arnold to look further into the matter.[21] By the time Arnold wrote back, Bernoulli had confirmed from his nephew that his letter had been delivered.[22] Bernoulli then wrote to De Moivre on March 20, 1714, expressing his concern that the dispute between Newton and Leibniz might be behind the cessation of their correspondence. Bernoulli wondered in his letter if De Moivre thought that Bernoulli and his nephew Nicolaus had taken Leibniz's side in the dispute. He assured De Moivre that they had a high regard for both Newton and Leibniz, and that he and his nephew were taking a neutral position with the hope that both sides would compromise a little.[23] Before receiving a reply from De Moivre, Bernoulli received a letter from John Arnold warning him to be careful about De Moivre—he was Newton's creature.[24]

As promised, De Moivre did send Johann Bernoulli a copy of the *Commercium Epistolicum*, as well as another one to Bernoulli's nephew, Nicolaus. It took a typical De Moivre route. He initially sent the books through Paul Vaillant, the Huguenot bookseller in London. From Vaillant's operation in The Hague, it went via the chaplain of the duc d'Aumont to Rémond de Montmort and then on to Bernoulli.[25] De Moivre informed only Nicolaus Bernoulli that the books were on their way. By the time the books arrived, they had received some damage from the rigors of eighteenth-century transport.[26]

De Moivre finally wrote to Johann Bernoulli in June 1714, explaining that he had been ill for some time; he had been "overwhelmed by headaches caused by the strain of travelling and teaching." In the same letter, he also described a mix-up with the distribution of the *Principia*, lamenting that as a result both he and Edmond Halley had to buy their own copies even though Newton had promised to present them each with one. He reassured Bernoulli that he did not think his correspondent had sided with Leibniz and then let slip with a comment, initially sugar-coated, that indicates that perhaps he did. De Moivre wrote:

> yet Sir, even if that was the case [siding with Leibniz], that would not have been reason enough for me to indicate dissatisfaction through my silence; indeed it would be utterly unreasonable if the gratitude I owe you for the kindness with which you have honoured me should vary due to such a minor incident; it does not seem that Mr. Newton was in the least troubled as to the distinction owed to the initial inventor of fluxions and differential calculus; he was in truth somewhat piqued by the insinuation of certain flysheets that he had learned the calculation of differences from Mr. Leibnitz and that some wanted to make him out as a plagiarizer. [27]

The flysheets De Moivre referred to was the *Charta Volans* that had been written (anonymously) by Leibniz in answer to the *Commercium Epistolicum*. Published in July 1713, it included an anonymized letter to Leibniz from Bernoulli that was critical of Newton's mathematical work and that hinted at Newton's reliance on Leibniz's results. The cloaking of the authorship of the letter was not done well enough so that many, in time, were able to guess that it came from the hand of Johann Bernoulli.

There is another hint of rebuke to Bernoulli in De Moivre's 1714 letter. Later in the letter, De Moivre referred to his own result on centripetal forces that he had sent to Bernoulli in 1705. He noticed that Bernoulli had published a paper in *Acta Eruditorum* in 1713 that contained the result and that, at the same time, gave credit of its discovery to De Moivre.[28] He thanked Bernoulli for this acknowledgment and told him that he had obtained some new results related to centripetal forces, which he had shown to Halley and Newton. He planned to publish the results in *Philosophical Transactions*. On the surface this appears all very nice and friendly. There is an unwritten subtext to all this. Bernoulli's paper in *Acta Eruditorum* points out Newton's error in Book II, Proposition 10 of the *Principia*, thus bringing into question Newton's ability as a mathematician.[29] By thanking Bernoulli for the acknowledgment explicitly, he also informed Bernoulli implicitly that he had carefully read the paper. Since De Moivre had informed Bernoulli in 1712 that Newton had corrected the error, this was, unstated by De Moivre, a serious faux pas on Bernoulli's part.[30] De Moivre had now learned the art of criticism through politesse within the Republic of Letters. Although Bernoulli responded to De Moivre's 1714 letter expressing hope for peace between Newton and Leibniz, De Moivre never wrote to Bernoulli again. A year and a half after Bernoulli's last letter to De Moivre, he again wrote to John Arnold asking him to look into whether De Moivre had received his letter.[31]

During 1713, Keill continued his attack on Leibniz by publishing an article anonymously in French in the May/June issue of *Journal literaire*.[32] Published out of The Hague by a Scotsman named Thomas Johnson, *Journal literaire* was the first literary journal in French to appear in the Low Countries. It was one of a number of books and journals intended for a French audience that circumvented France's printing and censorship laws. Keill's 1713 article went over Newton's discoveries in detail and reproduced the report to the Royal Society on Newton's priority to the discovery of the calculus. Leibniz responded to Keill in an article that appeared in the November/December issue of *Journal literaire*.[33] Also attached to Leibniz's response was a French translation of the *Charta Volans*.

Newton had already seen a copy of the *Charta Volans* in the autumn of 1713 and had planned to publish a response to it. He wrote several drafts of his response in English, addressing his response to Thomas Johnson, probably intending the response to appear in *Journal literaire*.[34] The first indication that De Moivre was becoming more involved in the dispute behind the scenes is that he translated one of Newton's drafts into French. Newton never submitted his letter to Johnson. Rather, it was Keill who took up the cudgel on his behalf, but with Newton standing directly behind him.[35] Keill published a lengthy article in the July/August issue of *Journal literaire* for 1714,[36] broadly hinting that Leibniz had got his ideas from Newton. Previous to the submission of his article, Keill expressed in a letter to Newton the hope that De Moivre would translate into French what he had written in English. The translation request went through Halley.[37] At the time, Keill was Savilian Professor of Astronomy at Oxford while Halley, De Moivre's close friend, was Savilian Professor of Geometry. Presumably De Moivre did the translation. It is possible but doubtful that De Moivre was involved earlier by providing Keill with a French translation of his 1713 article in *Journal literaire*. The doubt is based on Keill expressing in the same letter to Newton his lack of trust in those at *Journal literaire* to do a proper translation for his new article, indicating that perhaps the translation for the 1713 article was done in The Hague and was not to his satisfaction.

Keill continued his attacks even after Leibniz's death on November 14, 1716, and prior to that enlarged his range of attack to include Johann Bernoulli. The background to the attack on Bernoulli was research on central forces.[38] Keill had written a paper on the subject in 1708 and published it in *Philosophical Transactions*; the 1708 volume was not printed and distributed until 1710. In his paper, Keill used the formula on centripetal forces that De Moivre had discovered and communicated to Bernoulli in 1705. When he first obtained it, De Moivre had also sent the formula to Keill and Halley at Oxford, as well as to David Gregory.[39] In *Journal literaire* for 1716,[40] Keill claimed that there was nothing new in a 1710 letter that Bernoulli had published in the *Histoire de l'académie royale des sciences*. Everything Bernoulli had done, Newton had already done in the *Principia*, Keill claimed. Moreover, Keill had already obtained Bernoulli's result on central forces in 1708. Bernoulli's letter is the one he had written to Jacob Hermann on centripetal forces and subsequently read before the Académie royale des sciences in 1710. Once he finished with the *Principia*, Keill went on to take up De Moivre's cause. He pointed out that the theorem on centripetal forces that Bernoulli mentioned in the letter, which Bernoulli claimed he had obtained in 1706 and communicated to De Moivre,[41] was actually first communicated in the opposite direction. Keill noticed that Bernoulli had mentioned the result in his *Acta Eruditorum* article in 1713 and had credited De Moivre there. Why, Keill wondered, had Bernoulli hidden this in 1710? Bernoulli replied, or had someone else reply, to Keill in an anonymous letter in the *Acta Eruditorum* for 1716. Keill hammered back in a letter to Bernoulli the next year and published it in *Journal literaire* in 1719.[42] Keill's letter went over the whole history of the calculus dispute and, after several pages, he began to taunt Bernoulli. He was surprised that Bernoulli could not recognize that his results on central forces were the same as Newton's and went on to accuse Bernoulli of plagiarism with respect to Keill's work on central forces. He compared the situation to De Moivre's; Bernoulli could have claimed priority for De Moivre's theorem on centripetal forces, Keill said, if De Moivre had not been alive to produce epistolary evidence to the contrary. Whether all this was done, both in 1716 and 1719, with De Moivre's knowledge and approval, is difficult to say. If De Moivre continued to do French translations for Keill, then the answer can only be in the affirmative. All that comes to us in the surviving historical evidence is that in 1718 from his chateau in France, Montmort believed that De Moivre was behind Keill's letters.[43] That may not have been a well-founded belief in view of Montmort's fixation on De Moivre's alleged plagiarism.

There is one instance in which De Moivre definitely did provide a French translation during the calculus dispute. It was for Brook Taylor. In 1715 Taylor published his *Methodus Incrementorum*, which today is famous for its section on the development of the Taylor series expansion in mathematics.[44] The book is tersely written and contains references only to Newton's work. One of the propositions in the *Methodus* contains Taylor's results on the center of oscillation in compound pendulums.[45] At about the time that Taylor was working on this, Johann Bernoulli was also working and publishing on the same problem. The July 1716 issue of *Acta Eruditorum* contains an anonymous unfavorable review, written by Leibniz, of the *Methodus* followed by an anonymous letter written by Johann Bernoulli.[46] The letter outlines Bernoulli's contributions to the development of the calculus, as well as a description of Newton's shortcomings and criticisms of Keill. Inserted in the middle of the letter are some general accusations against Taylor, and some others, of plagiarizing Bernoulli's work. Taylor wanted to respond, but since the *Acta Eruditorum* was taking Leibniz's side in the calculus dispute, he needed another venue in which to respond. He chose *Bibliothèque angloise*. Published in Amsterdam, this was a literary journal written and edited by Michel de la Roche, De Moivre's friend and former mathematics student. Part of the journal's purpose was to disseminate Newtonian science throughout the Continent by including regular reviews of English scientific books. Taylor anonymously wrote his own review of the *Methodus Incrementorum* for *Bibliothèque angloise*.[47] The review included a response to Bernoulli. Originally written in English and strongly worded in places, De Moivre dutifully translated the review into French with one word change in one sentence to tone it down ever so slightly. In a letter dated November 2, 1717, De Moivre wrote to Taylor with the suggested word change and informed him that de la Roche was afraid to publish the review, "thinking that he would be called to account for some of the expressions that are in it."[48] The review did appear in 1718, probably with some serious editing by de la Roche. The offending sentence that De Moivre reworded for Taylor and quoted back to him in his letter, does not appear in the review.

There were two individuals, one French and the other Italian, who tried to be peacemakers in the calculus dispute. The Italian, Abbé Antonio Schinella Conti, was in London between 1715 and 1718. He had come to London specifically to meet Newton and soon became part of Newton's intellectual and social circle. Previously, he had been in correspondence with Leibniz.[49] Conti also had access to the court and often met with Caroline, Princess of Wales, who for several years had been in close contact with Leibniz in Hanover and earlier in Berlin.[50] The Frenchman, Pierre Varignon, was director of the Académie royale des sciences between 1711 and 1719; he had been a member of the Académie royale since 1688. In the 1690s, Varignon was one of those who introduced France to Leibnizian calculus. At the same time, after reading the *Principia*, he was a confirmed Newtonian in his scientific outlook. Varignon became a fellow of the Royal Society in 1714 (nominated by De Moivre's friend, William Jones) and Conti a year later (nominated by Newton). During his time in London, Conti was a go-between for Newton and Leibniz. From about 1719 until his death in 1722, Varignon acted in the same capacity for Newton and Bernoulli. Varignon was able to involve De Moivre in his efforts to achieve a reconciliation between the two mathematical giants.

De Moivre's involvement in the Leibniz-Conti-Newton triangle stems from a challenge problem set late in 1715 by Leibniz with input from Johann Bernoulli.[51] Knowing that the general solution could not be obtained by the geometrical methods used in the British approach to the calculus, the problem was meant to test the abilities of British mathematicians. Leibniz sent the problem to England via Conti, who circulated it. The problem in general is to find the orthogonal trajectories of any family of curves that is represented in a single equation. A curve that is an orthogonal trajectory crosses all curves in the family at a right angle. Leibniz inadvertently made the problem much easier by giving a specific case as an example—the family of hyperbolas with the same vertex and the same center. Except for Newton, the English mathematicians focused on the hyperbola rather than a general family of curves. The solid curves in the diagram are from the family of hyperbolas of the form $y^2/\alpha - x^2/4 = 1$, where α indexes the family. The lower curve has $\alpha = 1$, the

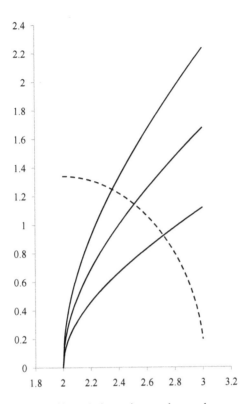

A family of hyperbolas with an orthogonal trajectory.

middle $\alpha = 1.5$, and the upper $\alpha = 2$. For this family of hyperbolas, the associated curves of orthogonal trajectories are of the form $y^2 + x^2 - 8\ln(x) = \beta$, where now β indexes this family. The dashed curve in the diagram is the trajectory with $\beta = 0.25$. When Bernoulli saw how Leibniz had expressed his challenge problem, he wrote to him in January 1716 explaining that the case of the hyperbola had an easy solution. Bernoulli concluded in his letter:

> Therefore, we must fear that the English Analysts will solve this case with common methods; and then, when they notice that they have succeeded so easily they will emerge even more puffed-up with pride and even more confirmed in their belief that they are superior.[52]

Many British mathematicians, De Moivre among them, solved the problem for the hyperbola. Newton's general solution, published anonymously, was not very successful.

De Moivre's solution to Leibniz's challenge problem has not survived. At the time when he saw it, John Machin thought De Moivre's solution "exceedingly neat and simple," so much so that he liked it better than his own solution. De Moivre also recognized the ambiguity in the statement of the problem and told Conti of his concerns. When Conti was about to report back to Leibniz, he showed his letter to both Newton and De Moivre before sending it on in March 1716. De Moivre made some corrections to the letter and inserted his concerns about the ambiguity of the problem as stated.[53]

With continued attacks on Johann Bernoulli after Leibniz's death, the calculus dispute simmered on. Beginning in 1718, Pierre Varignon initiated his efforts to bring about a reconciliation between Newton and Bernoulli. He made his first move based on receipt from Newton of three copies of the new edition of Newton's *Opticks* in English. The copies were sent to Varignon in early August of 1718. Varignon wrote to Newton saying that he could not read English well so he had lent a copy to an English friend; he planned to learn about any new results after his friend read the book. In the same letter, Varignon told Newton that he had sent a copy to Bernoulli on Newton's behalf "in order that I might reveal your generous nature to him." Two months later, in October, Newton sent Varignon five copies of the newly printed Latin edition of the *Opticks*.[54] Newton intended one of the copies for Varignon and specified the distribution of three of the remaining four copies. The last copy, Newton told Varignon, was for any of Varignon's friends who could understand the subject. Varignon took this opportunity to send the remaining copy to Bernoulli, telling Newton that it was done "for the sake of bringing peace and concord between you and him."[55]

A thaw had begun but it was not an easy one—for example, the attack by Keill on Bernoulli that appears in 1719 did not help warm Bernoulli to Newton and his supporters. Varignon enlisted De Moivre's help in the peacemaking process. He had been corresponding with De Moivre since at least 1707 and the two were on good terms.[56] He was also probably well aware of De Moivre's closeness to Newton. Unfortunately, none of the De Moivre–Varignon correspondence is extant. The only reference to it appears in surviving letters between Varignon and Johann Bernoulli.[57]

Johann Bernoulli did write to Newton on June 24, 1719, to thank him for the new English and Latin editions of the *Opticks*. He remained wary of Newton and did not trust several of the British mathematicians: Keill had attacked him in print several times; he thought Brook Taylor had plagiarized his work; and Bernoulli suspected De Moivre of helping Keill, in addition to ceasing their correspondence five years earlier. Despite this, he still wanted a renewed friendship with Newton. Lying outright, he assured Newton that he was not the author of the letter to Leibniz that had appeared in the *Charta Volans*. Newton, equally as wary, wrote back on September 29, 1719. After referring to the letter in the *Charta Volans* and saying that he now assumed that Bernoulli was not the author, Newton expressed his desire for friendship and concluded, "I shall make it my business to settle the arguments which you have with my friends, as far as in me lies." Bernoulli wrote again to Newton on December 10, 1719, and this time De Moivre reported to Varignon on Newton's reaction to it. At first Newton was pleased with the letter but later he became angered by it. Newton drafted a reply to Bernoulli but never sent it. Varignon tried for a reconciliation one more time before his death in 1722 by sending Bernoulli three "elegantly bound" copies of the new French edition of Newton's *Opticks* that he had received from Newton. Bernoulli wrote to Newton on January 26, 1723, thanking him for the copies. He also made reference to his letter to Leibniz that had been published in the *Charta Volans*. It had recently been reprinted in 1720 in a publication entitled *Recueil des diverses pièces*.[58] Bernoulli admitted to authorship of the letter but denied that he had any hand in certain additions to it, namely praising himself as an excellent mathematician.[59] De Moivre was involved in the French edition of the *Opticks* and, perhaps in some way, in the *Recueil des diverses pièces*.

The two-volume *Recueil des diverses pièces* was compiled and edited by De Moivre's old friend Pierre Des Maizeaux. Des Maizeaux had collected several contemporary letters and papers concerning the calculus dispute, prefaced it with a history of the dispute that relied on the *Commercium Epistolicum*, and published it in 1720 in French out of Amsterdam. The *Recueil* was not kind to Johann Bernoulli; it referred unfavorably to Bernoulli's anonymous letter to Leibniz published in the *Charta Volans*. A letter from Newton to Conti was included in the *Recueil* in which Newton used the phrase "alleged mathematician" (*prétendu mathematicien*) when referring to the author of the anonymous letter. Later in the *Recueil*, Newton commented to Conti that, in view of Bernoulli's actions, mathematics in the future would become acts of knight-errantry instead of reason and demonstration. The inclusion of these letters with their offending phrases, of course, infuriated Bernoulli.

Newton knew of Des Maizeaux's work well prior to its publication. Des Maizeaux had sent Newton proof-sheets of the *Recueil* in 1718 to which Newton made corrections.[60] The publication of the *Recueil* was delayed initially because Des Maizeaux added new material. Newton may also have had a hand in the delay once he read of Bernoulli's claim that he was not the author of the letter to Leibniz. Even before he saw it, Bernoulli was angered by the prospect of the new publication and felt that Newton was behind it all. Once published, Des Maizeaux sent Varignon a copy of the *Recueil*. Varignon wrote to Des Maizeaux and De Moivre about the difficulty that the publication was making in achieving a reconciliation.[61]

It is difficult to say exactly what role De Moivre played in the *Recueil*. De Moivre and Des Maizeaux had been friends for many years, meeting regularly at the Rainbow Coffeehouse probably well before Newton and De Moivre began meeting at Slaughter's Coffeehouse. Newton's knowledge of French has been described as "sketchy."[62] For example, when he began working on alchemy, Newton probably had help from Fatio de Duillier in order to read and digest the alchemical literature that was in French. When he wrote to Varignon, it was typically in Latin, not French. Consequently, Newton may have had help from De Moivre as he went through the proof-sheets of the *Recueil*. Whatever role De Moivre did play, Varignon was still looking to De Moivre for help in the peacemaking process. The phrases "alleged mathematician" and "knight-errant" remained the sticking points. These phrases, appearing in letters from Newton to Conti, may have been late additions to the *Recueil* and not seen by Newton.

A letter from Varignon to Bernoulli dated February 15, 1721,[63] which quotes at length a letter from De Moivre to Varignon, shows the extent of the efforts De Moivre was making on the English side in the attempted reconciliation. De Moivre reported to Varignon that Newton had chastised Des Maizeaux for stirring the pot with material on Bernoulli in the *Recueil* and making Newton look bad. Afterward, Des Maizeaux came to his friend De Moivre for advice on what to do. De Moivre put him off for a few days so he could talk to Newton about it without telling Des Maizeaux of his intentions. Subsequently, De Moivre went to talk to Newton and worked out a scheme to smooth things over between Newton and Bernoulli through Varignon. De Moivre's letter to Varignon was written before Des Maizeaux was brought back into the picture. All De Moivre said to Varignon was that he told Des Maizeaux, "we would be seeing each other at our pleasure and that we would speak on the matter again."[64] Presumably, that would have been some future evening at the Rainbow Coffeehouse.

Varignon's final attempt at a reconciliation in 1722 before his death that year was to draft a letter stating that Newton had no animosity toward Bernoulli. Through De Moivre, it received Newton's approval for publication. However, Bernoulli would not approve it for publication and made further demands. At that point, Varignon gave up hope of any reconciliation and died soon thereafter.

Although he was unsuccessful in bringing Newton and Bernoulli to a reconciliation, Varignon was successful in bringing to fruition a project of interest to Newton that also involved De Moivre. Pierre Coste, another of De Moivre's old friends from the Rainbow Coffeehouse, had translated Newton's *Opticks* into French. It was published in Amsterdam in 1720. As it stood with the French printing and censorship laws, it was illegal to import the Amsterdam edition into France. A French publisher wanted to bring out an edition of Coste's translation in Paris. Since the book was a scientific one, the process of approval went through the Académie royale des sciences. This eventually involved Varignon, who wrote to Newton about the project in May 1720.[65] Subsequently, Varignon took charge of getting the book to press.

The new French edition of the *Opticks* led to a little tension between some old friends.[66] When Newton wrote to Varignon in January 1721 thanking him for undertaking the new edition, he told Varignon that De Moivre would be sending him corrections to the French edition. There was no mention of Coste, the original translator. Newton also told Varignon to ignore any corrections that might come from others. If there were any other corrections, they would come through Newton. By August 1721 it is apparent in Newton's letters to Varignon that De Moivre was closely involved in the editorial process from the English side. That same month, Coste sent his own corrections to Newton, as well as comments on some of De Moivre's corrections, complaining that he had not seen all of De Moivre's corrections and had been treated badly. A few days later Coste informed Varignon in a letter that Newton had promised to send on his corrections. Varignon subsequently received the corrections and in September Newton wrote to Varignon that he should choose whatever corrections he thought best. Coste does not seem to have made any corrections to the mathematical material, while De Moivre did. Varignon wrote to Newton in July 1722 saying that De Moivre's corrections to the mathematical material were very helpful. He also noted that when De Moivre's and Coste's corrections overlapped, they were often identical. The printing was finished by October 1722. The two old friends, Coste and De Moivre, seem to have reconciled fairly quickly. When tensions were building in August 1721 over the corrections to the *Opticks*, Coste visited De Moivre at his lodgings. Coste came out apparently satisfied with the way in which Varignon would handle the corrections and promised to acknowledge De Moivre's work in the book's preface. When the book appeared, the preface did indeed contain high praise for De Moivre's contributions.[67]

There is story about Newton late in life, which is probably apocryphal, but which shows De Moivre's status as a mathematician and his close relationship to Newton.

> It is reported that, during the last ten or twelve years of Newton's life, when any person came to ask him for an explanation of any part of his works, he used to say: "Go to Mr De Moivre; he knows all these things better than I do."[68]

When Newton died in 1727, his nephew by marriage John Conduitt decided to write a biography of him. Conduitt solicited information from several of Newton's friends and colleagues. De Moivre came forward with a number of details of Newton's early life and work, events that occurred well before De Moivre and Newton met.[69] The information must have accumulated in De Moivre's memory from the several evenings over the years they had spent together at Slaughter's Coffeehouse.

❧ 7 ❧

Miscellanea Mathematica

Throughout all the Newtonian distractions of the calculus dispute and the demands of his teaching, De Moivre continued to work on his own mathematical research that was carried out along four broad, and sometimes interrelated, lines. His results in celestial mechanics, which began with his work on centripetal forces in 1705, saw some further development. Although the publication, finally, of Newton's early work on quadrature was far-reaching with seemingly little left to do in the area, De Moivre carried out some additional minor work on the quadrature of a particular curve. With hints of it in *De Mensura Sortis* and certainly with his early work anticipating Taylor series expansions that he had communicated to Johann Bernoulli in 1708, he began to expand his interests in the area of infinite series. And, of course, his interests in the theory of probability developed more deeply.

The paper on centripetal forces mentioned to Bernoulli by De Moivre in mid-1714 was not presented to the Royal Society until December 20, 1716.[1] It was printed in 1717 in *Philosophical Transactions*.[2] Two general theorems on central forces were set out in the paper: his own from 1705 and one due to Newton from the *Principia* which states that the centripetal force exerted on a body travelling along a curve that is a conic section is inversely proportional to the square of the distance of the body to the focal point of the conic section. From these two general theorems De Moivre determined results on the velocities of bodies along their trajectories in conic sections.

One result he obtained harks back to 1705 when it was probably Thomas Sprat, Bishop of Rochester, who asked De Moivre a question about comets. De Moivre obtained for the bishop the ratio of the velocity of a comet (in a parabolic orbit shown as a dashed line in the diagram) to the velocity of Earth (in a circular

A conflation of De Moivre's diagrams from his note to Thomas Sprat in 1705.

orbit in the diagram) both travelling about the sun, which is the focal point of the two orbits.[3] Corollary 4 in the 1717 paper finds the ratio of the velocities of any two bodies travelling along conic sections with the same focal points.

The next year, on March 20, 1718, De Moivre presented another paper to the Royal Society on celestial mechanics. A description of the paper's contents is:

> A paper of Mr. Moivre's concerning the maxima that occurs in planetary motions as in what points the planets recede swiftest from the sun and where their real and angular velocities alter swiftness &c was produced and ordered to be printed.[4]

The paper appears in *Philosophical Transactions* for 1719.[5]

The two papers on celestial mechanics are the foundation of the last book, Book VIII, of De Moivre's *Miscellanea Analytica* published in 1730. Book VIII is devoted to centripetal forces; it contains some repetition of material in the 1717 paper, followed by a general treatment of maxima and minima in the motions of celestial bodies, more general than what appears in the 1719 paper. De Moivre's work on centripetal forces had some impact. The results became part of John Keill's lectures on astronomy at Oxford. The lectures were published in 1721.[6]

During this time period, De Moivre also completed some minor work on curves. The published work has the appearance of an extract of correspondence with someone else, unnamed.[7] There is no record of its presentation before a meeting of the Royal Society. Published in 1715, what De Moivre did was to study the curve given by the equation $y^3 + y^2x + yx^2 + x^3 = axy$, where a is a constant. This is related to the folium curve proposed by René Descartes more than seventy-five years earlier. The folium curve is defined by the equation $y^3 + x^3 = 3axy$, again where a is a constant. An example of De Moivre's curve is shown in the diagram. The curve gets its name, folium or foliate, from the leaf shape evident from the enclosed area of the curve. As noted by De Moivre, his curve resembles curve number 40 in Newton's catalog of third order curves. These types of curves, like those of De Moivre and

Descartes, have terms in x and y such that in any one term the sum of the powers, i.e., $u + v$ in the product $y^u x^v$, is less than or equal to three. The catalog is part of Newton's manuscript on third order curves, *Enumeratio linearum tertii ordinis*, which was published in both the *Opticks* in 1704 and *Analysis per Quantitatum Series, Fluxiones, ac Differentias* in 1711.[8] Among other properties, De Moivre was able to obtain the area within the enclosed part of the curve shown in the diagram.

Pierre Rémond de Montmort visited England in 1715 to meet with other mathematicians and scientists, and to observe the solar eclipse that occurred on May 3, 1715. Montmort watched the solar eclipse with Edmond Halley and some others.[9] At other times during his trip, he met with both Brook Taylor and Abraham De Moivre. De Moivre says that he acted as Montmort's interpreter and guide at times.[10] The same year that Montmort travelled to London, Brook Taylor's *Methodus Incrementorum* appeared in print. In the book, Taylor applied his newly developed methods to a variety of problems including the valuations of sums of infinite arithmetical series. An example of such a problem would be finding the numerical value of

$$\frac{1}{1 \cdot 2 \cdot 3} + \frac{1}{2 \cdot 3 \cdot 4} + \frac{1}{3 \cdot 4 \cdot 5} + \frac{1}{4 \cdot 5 \cdot 6} + \cdots.$$

Montmort became interested in infinite series problems, probably as a result of his trip to England. After his return to France, he wrote Taylor about five or six

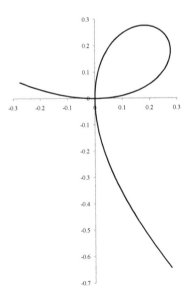

De Moivre's curve with $a = 1$.

letters over the years 1715 to 1717 that contain his solutions to some infinite series problems.[11] Many of the problems he considered and wrote about in French to Taylor appear in a paper in *Philosophical Transactions* in Latin.[12] The paper is long and contains several lemmas and propositions that are required to obtain the appropriate series sums.

In October 1718 De Moivre received a letter, which is no longer extant, from Brook Taylor about problems in infinite series and how they could or could not be solved using the techniques in *Methodus Incrementorum*. From De Moivre's reply, it is obvious that he had read Montmort's paper in *Philosophical Transactions*. In his reply, De Moivre outlined a method of summing infinite series that uses a clever mathematical trick rather than Taylor's methods in the *Methodus* or Montmort's more complicated approach.[13] De Moivre did admit in the letter that, despite the power of his own approach, Taylor's methods were more generally applicable than his.

To illustrate De Moivre's trick, consider the infinite series

$$z = 1 + \frac{x}{2} + \frac{x^2}{3} + \frac{x^3}{4} + \frac{x^4}{5} \cdots ,$$

where x is some number. According to De Moivre, it does not matter whether the series under consideration converges or not. De Moivre uses the word *summable* instead of convergent. Take the series in x, and multiply the left side of the equation by $(x-1)^2$; then the right side by its equivalent expression, $x^2 - 2x + 1$. After collecting terms in powers of x on the right side of the equation, the whole equation can be expressed as

$$(x-1)^2 z = 1 - \frac{3}{2}x + \frac{2}{1 \cdot 2 \cdot 3}x^2 + \frac{2}{2 \cdot 3 \cdot 4}x^3 + \frac{2}{3 \cdot 4 \cdot 5}x^4 + \cdots .$$

On setting $x = 1$ the equation becomes

$$0 = -\frac{1}{2} + \frac{2}{1 \cdot 2 \cdot 3} + \frac{2}{2 \cdot 3 \cdot 4} + \frac{2}{3 \cdot 4 \cdot 5} + \cdots$$

so that the original infinite series

$$\frac{1}{1 \cdot 2 \cdot 3} + \frac{1}{2 \cdot 3 \cdot 4} + \frac{1}{3 \cdot 4 \cdot 5} + \frac{1}{4 \cdot 5 \cdot 6} + \cdots$$

sums to $1/4$. De Moivre had said that the method would work even if the original series in x does not converge, which in this case it does not. The series in x is the expansion of $1 - \ln(1 - x)$, which does not converge at $x = 1$. A little calculation shows that the numerical series does converge to 1/4, albeit slowly. De Moivre's general method, illustrated here only by this simple numerical example, appears thirteen years later in Book VI on series in De Moivre's *Miscellanea Analytica*.

After his general exposition that included this simple numerical example, De Moivre concluded his letter to Taylor by writing, "Do me ye favour not to mention anything of this Method to Mr Monmort. You may, if you think, to let him know that I can do it easily." These two sentences put in a nutshell how De Moivre approached Montmort between 1715 and Montmort's death in 1719. It also shows that despite Taylor's apparent closeness to Montmort, De Moivre trusted Taylor not to reveal any of De Moivre's new mathematical discoveries to Montmort.

Up to and including the time of Montmort's visit to London, De Moivre has described his relationship with Montmort. De Moivre claimed to have put aside what was to him the offending letter from Montmort to Nicolaus Bernoulli that appeared in the second edition of *Essay d'analyse*. After Montmort's book appeared, De Moivre began corresponding with him about problems concerning summation of series.[14] After Montmort's visit, De Moivre claimed that they were both too busy to correspond with one another. De Moivre was preparing his *Doctrine of Chances* for publication; Montmort was too busy working on his research in infinite series that appears in the *Philosophical Transactions*. This does not ring true. From 1715 until his death, Montmort carried on a substantial correspondence with Brook Taylor as well as with both Johann and Nicolaus Bernoulli. After meeting face-to-face in London, for some reason De Moivre must have developed a dislike for Montmort and stopped communicating with him after Montmort returned to France.

De Moivre's researches into probability also continued. He received encouragement in his work after receiving a visitor from France in 1713. Soon after the signing of the Treaty of Utrecht that ended the War of the Spanish Succession, André-François Deslandes arrived in London in the entourage of the French ambassador Louis de Villequier, duc d'Aumont. The year before in France, Deslandes obtained an entry-level mathematics position (*élève géomètre*) at the Académie royale des sciences in Paris. While in London, Deslandes dined at Newton's home with Halley, De Moivre, and Craig as additional guests.[15] After meeting De Moivre, Deslandes wrote the following poem in his honor.[16]

To Abraham De Moivre
Eminent teacher of mathematics

O new Euclid, to which noble man has bounteous Nature revealed her mysterious secrets?

Come on! Break off your delay and search your eloquent mind.

Why do you wish to be concealed? Why do you cover up your learning?

It is wrong and a crime to have knowledge for yourself alone;

dare to have knowledge for others,[17]

and see to it that a work worthy of your name soon shines
and that it is polished with the art of a learned mind.
You, skilful and eager, a clear-sighted lover of truth,
have illuminated the difficult paths of hidden places and the secret of Algebra.
In your wisdom you unite the austere Archimedes with the very tender Catullus.
Therefore, whether you undertake friendly lessons with agreeable poems of nature,
or majestic and sacred charms of wisdom:
we will look upon you with wonder and, joyously,
we shall celebrate your talents in golden songs.

The line that includes illuminating "the difficult paths of hidden places and the secret of Algebra" is probably a reference to the recently published *De Mensura Sortis*. De Moivre may have been working on new results in probability already and then talked about his work to Deslandes. Or, he may have been speculating about what he was going to do next in probability—hence the admonishment not to keep knowledge to himself. Coupling the "austere Archimedes" with the "very tender Catallus" is probably a reference to De Moivre's mathematical abilities combined with his interests in classical French literature, especially the works of Rabelais and Molière.

After Nicolaus Bernoulli visited London in 1712, he kept up his contact with De Moivre through correspondence. When his uncle Jacob's posthumous book[18] *Ars Conjectandi* finally saw print in 1713 (Jacob Bernoulli died in 1705), Nicolaus intended to send copies of it to De Moivre, as well as to Newton and Halley.[19] Nicolaus had written a forward to the *Ars Conjectandi* in which he noted that his uncle's work on the economic and political applications of probability was incomplete. Jacob had been in poor health and had died before completing his manuscript on that material. In the forward, Nicolaus encouraged De Moivre and Montmort to take up where his uncle had left off in these two areas of application. By the time Nicolaus's letter with the offer of sending the book reached De Moivre (it was sent December 30, 1713), he had already bought a copy of the book among four that had recently reached the booksellers' stalls in London. When De Moivre wrote back on March 3, 1714, he thanked Bernoulli for his mention in the forward. He said that he would like to work in the suggested areas of application but that his students took up all his time.[20]

Bernoulli's 1713 letter to De Moivre contains his general solution to Waldegrave's problem, or the problem of the pool; he also sent a copy of his solution to William Burnet.[21] De Moivre had solved the problem for three players in *De Mensura Sortis*, but his generalization to more than three players, as pointed out in Montmort's September 5, 1712, letter to Nicolaus Bernoulli, was impracticable.[22] De Moivre liked Bernoulli's new solution and presented his paper on it to a meeting of the Royal Society on February 11, 1714.[23] Later that same meeting, and probably at De Moivre's instigation, Newton proposed Nicolaus Bernoulli as a candidate for fellowship in the Royal Society; he was elected a month later on March 11. Bernoulli's general solution to Waldegrave's problem appears in the issue of *Philosophical Transactions* for last quarter of 1714.[24] The article is followed immediately by an article containing De Moivre's solution to the problem for four players using his original approach (which Montmort had claimed to be impracticable) and an indication of how to proceed for six players.[25] Anders Hald, confirming Montmort's original assessment, has commented that De Moivre's "method is theoretically simple but in practice very cumbersome for more than three players."[26]

Bernoulli wrote again to De Moivre on August 4, 1714. De Moivre had told Bernoulli that all probability problems in games of chance could be solved using combinations or by series expansions. In the letter, Bernoulli challenged this claim by setting the following problem for De Moivre to solve. Two players, A and B, respectively, put a and b tokens or units of money into a pot. Then they play with a four-sided die with faces marked 0, 1, 2, and 3. With one twist to the game, the player throwing the die takes from the pot the number of tokens corresponding the value on the face that shows. The twist is that when player A throws a 0, he puts one token into the pot. If he throws more than 0 and the number of tokens in the pot is less than the number thrown on the die, then A puts the difference into the pot. The problem is to find the initial numbers of tokens a and b that must be put into the pot so that the two players, A and B, have equal chances to win the pot. De Moivre did not respond to the challenge and, furthermore, never wrote to Bernoulli again.[27]

Essentially, the same thing happened to Montmort. When Montmort was in London, he met De Moivre. Montmort's impression from their meeting was that his relationship with De Moivre was very friendly. After returning to France, he tried to begin a correspondence with De Moivre. In 1716, Montmort heard from Newton that De Moivre was planning a new edition of his work on probability.[28] The new book would contain "considerable improvements." Upon hearing that De Moivre was planning this time to write in English, he wrote to him expressing the hope that the work would be in Latin, feeling that Latin would give the work a wider readership than English. At the same time, Montmort sent De Moivre ten theorems that he felt could be included in his new publication. Even after receiving two letters from Montmort, De Moivre did not reply. Concerned about the status of his ten theorems, Montmort wrote to Brook Taylor in April 1716, asking Taylor to look into the matter discreetly.[29] Three months later in July, Montmort wrote again to Taylor saying he had received nothing from De Moivre for the past eight months and asking about the publication status of De Moivre's book on probability.[30] Montmort tried sporadically to engage De Moivre. He wrote to Brook Taylor on October 17, 1717; still under the Julian calendar, Taylor wrote at the top of the letter that it was received on October 16. The postscript to the letter contains a challenge problem set by Montmort for De Moivre to consider; Montmort told Taylor he had already solved the problem.[31] Taylor passed the problem on to De Moivre. Replying on November 7, 1717, to one of Taylor's earlier letters to him, De Moivre said that, "I have not yet undertaken ye Problem of Mr Montmort...."[32]

De Moivre told Taylor that the problem was similar to analyzing a game that is now known as peg solitaire, a board game that was fashionable in France in the court of Louis XIV. He gave some ideas of how to approach the problem but went no further. In the same letter, De Moivre told Taylor that he was very busy working on his book.

When Johann Bernoulli first heard from Montmort in 1716 of De Moivre's plans to write his book on probability, he told Montmort that he hoped De Moivre would send him a copy of it.[33] Two years later, in 1718, when he heard that the book was in English, he wrote to Montmort asking why the book was written in that language. Had De Moivre forgotten his French or Latin, Bernoulli wondered? He now saw De Moivre as totally anglicized; and if De Moivre did not send him a copy of the book, he would take it as a sign that he had fallen totally from De Moivre's favor.[34]

De Moivre stopped writing to both Bernoullis by mid-1714 and seems not to have written to Montmort after Montmort's 1715 trip to London. There are several possible reasons for this. Although it did not stop him from helping Newton, Keill, Taylor, and perhaps other British mathematicians, he was busy making his living at tutoring. His self-imposed break with the Bernoullis might be connected to the calculus dispute. He may have been offended by Montmort's apparent condescending attitude toward him and his work. In addition, Nicolaus Bernoulli and Montmort were both in competition with him in terms of developing probability theory. We have already seen De Moivre asking Taylor not to reveal his work on series to Montmort. There are other possible reasons beyond professional relationships. De Moivre had written to Johann Bernoulli in 1714 that he had been suffering from headaches for several months. Even Montmort had heard, perhaps from one of the Bernoullis, that De Moivre had been sick and expressed his concern to Taylor in a letter of January 2, 1715.[35] De Moivre had been ill before this time. In his 1705 note to Thomas Sprat, he apologized for not writing to Sprat sooner, but he had been "a little indisposed." Three years before that, he had suffered from a bout of smallpox. In a letter to Taylor dated September 29, 1718, there are also broad hints of De Moivre's health problems as well as allusions to another possible reason. De Moivre wrote:[36]

> I am sorry to hear that you are obliged to take ye same caution as I do in respect to health, to forbear ye reading of anything that requires too much application of thought. However I dont know but the in ye main, t'is best to make Mathematics a subject of Diversion, and not to be obstinately bent in unfolding ye Train of Thoughts of another Man, to which he has been led by degrees, and very often by Chance.

Perhaps he was just tired of deciphering other people's work. Whatever sickness he suffered from, De Moivre survived another thirty-six years.

There was another instance at this time in which De Moivre held back new mathematical developments from Montmort, and from several other mathematicians as well. In the third week of March 1718, De Moivre delivered a packet to his friend

Isaac Newton, possibly over coffee and conversation at Slaughter's, possibly at Newton's house nearby, or possibly just before a Royal Society meeting. As the first order of business at the meeting of the Royal Society on March 22, 1718, the president, Newton

> produced a paper sealed up with three seals intitled Mr De Moivres Demonstration of Some Theorems in his Book of Chances and the President informed the Society that Mr De Moivre desired the Society would make an entry in their minutes that such a paper was left in his custody.[37]

A week later, on March 29, an advertisement was placed in the newspapers informing subscribers to the book that they could pick up their copy of *Doctrine of Chances* beginning on Monday, April 7, at Slaughter's Coffeehouse between 4:00 and 8:00 p.m.[38] Whatever was in the Newton's possession, it must have been important to De Moivre—sealed with three seals, no less, when one was the norm on a letter.

❧ 8 ❧

The Doctrine of Chances
and the Doctrine Disputed

The packet in Newton's possession contained the outline of a new proof to the solution to the duration of play problem. When all the details are considered, it is a long and difficult proof that requires the development of a new mathematical theory of recurring series to get things started. It ends with some trigonometric arguments such that the numerical solution to the duration of play problem can be found easily using tables of sines or cosines. The trigonometric part of the solution harks back to his 1707 paper, with the explanation of it delayed until 1722, in which trigonometric methods are used to find the roots of a certain polynomial equation.[1]

De Moivre's solution is a direct response to comments from Montmort in the second edition of his *Essay d'analyse*, published in 1713.[2] There, Montmort published his correspondence with Nicolaus Bernoulli on the duration of play problem among others. Initially, Montmort obtained the solution to the duration of play problem when the two players have the same skill or the same chance to win a unit of money from the other player in any game. Bernoulli was able to get a very general solution for players with unequal skills. Montmort's comment was that his and Bernoulli's solutions were simpler and easier to calculate than the algorithm De Moivre used in *De Mensura Sortis*. With general accusations by Montmort of plagiarism on De Moivre's part in his *De Mensura Sortis*, De Moivre undoubtedly wanted this new proof to be considered entirely of his own invention. When he announced the new solution to the duration of play problem in *Doctrine of Chances*, he wrote in the preface:

> I hope the Reader will excuse my not giving the Demonstrations of some few things relating to this Subject, especially … the Method of Approximation

contained in page 149 and 150; whereby the Duration of Play is easily determined with the help of a Table of Natural Sines: Those Demonstrations are omitted purposely to given an occasion to the Reader to exercise his own Ingenuity. In the mean Time, I have deposited them with the Royal Society, in order to be Published when it shall be thought Requisite.[3]

In terms of facility of calculation, De Moivre's new method was much simpler than anything that appeared in *Essay d'analyse* on this subject. The requisite time, if De Moivre's actions are followed closely, was an interestingly convenient one. De Moivre had the packet opened before the Royal Society on May 5, 1720.[4] Montmort died about five months earlier on October 7, 1719, and thus was unable to lay claim to any part of De Moivre's new proof. The gap between Montmort's death and the opening of the packet gave De Moivre time to complete a manuscript that put flesh on the bones that were in the packet.

De Moivre brought to the meeting of May 5 a manuscript that enlarged on the material in the packet, which had been in Newton's possession since March 22, 1718.[5] The difference between the two is described in the minutes of the meeting: "The Paper containing the bare rules which are explained and proved at large in the Manuscript." De Moivre asked that the secretary of the Royal Society be given the two items in order to have him certify that the substance of the manuscript and the packet were the same. The secretary was Halley. He gave the material to two reviewers who reported back to him. The reports are still kept in the Royal Society's archives.

It is useful to quote in its entirety the first reviewer's report and to quote a short paragraph from the second reviewer.[6] Here is what the first reviewer writes:

> Having by order of the Society perused & compared M[r] De Moivres papers produced before the Society the 5th of May last, I find that the grounds & principles upon w[ch] the Propositions contained in the manuscript w[ch] he then produced, are built, are wholly contained in the papers w[ch] he deliver'd in to the Presidents custody on the 22 of May 1718 & which had remaind seald up in his custody until the said 5th of May last.
>
> The propositions in these papers seem to be the principles from whence he drew some other propositions printed in his Book of Chances without demonstration.
>
> And they contain a Method of summing up a various number of Infinite progressions w[ch] are not to be sum'd directly by any rules yet laid down by Writers on these subjects.
>
> Most of the Propositions seem to derived from this Observation. That If any quantity be divided by a Trinomial quantity & be converted into a series of an Infinite number of terms, each term in this series will be related to the two terms which immediately precede it, that is will always be a certain multiple

of one added to a certain multiple of the other. So that every term will always be necessarily determin'd by having the two terms wch immediately precede it.

Again if a quantity be divided by a Quadrinomial the several terms in this series will in like manner necessarily depend upon the three preceding ones. A quinquinomial on the 4 preceding & so on ad infinitum.

This being laid down. He solves this Problem.

A series of an infinite number of terms being given if each term be related to any number of preceding ones in the manner before expresst & the relation be known .. that is what the several multiples of the each several preceding terms are to make up any given term he then finds the Quantity from whence this series is generated, or which is the same thing he finds a certain number of Geometrical progressions the sum of all which are equal to the series first propos'd & vice versa.

Prob. 1   Two 3, 4 or more Geometrical progressions being given to find the relation of the several terms in a series consisting of the sum of each corresponding term of the Geometrical progression.

Prob. 2 To divide a series the relation of whose terms is given into several Geometrical Progressions the sum of whose terms make the series proposed.

Prob. 3 Two series of Terms of a given relation to the two preceding being given to find the relation of terms in a series form from the multiplication of the respective terms of the given series. See pag. 134 & pag. 154.

Prob. 4 The Division of the Circle applied to the Prob. for determining the Number of Games in which one of two gamesters will lose a certain number of Stakes.

Prob. 5 A series whose terms are related in a certain manner being given to find the sum of the terms taken at any equal intervals

Note A series arising from the Division of any Multinomial will have its terms with a given relation.

The second reviewer was quite terse. His short report does contain a paragraph that highlights how much savings in time could be gained from De Moivre's solution over those of Montmort and Bernoulli.

By help of this Method, he is able to solve some very difficult Problems in a very short time ye solution of which he conceives to be impracticable by any other Method, as an Instance of which he Proposes this Problem, A & B playing together till 45 stakes are won or lost of either side to find ye Odds of ye Play ending in 1597 Games, and which by any other Method would require ye addition of near 800 terms of a series.

In his *Doctrine of Chances*, De Moivre used as an example only four stakes, or units of capital, with play ending in forty games.[7]

The manuscript was published in a 1722 issue of *Philosophical Transactions*.[8] The delay in publication may have been due to the change in the editorship of *Philosophical Transactions*. Halley handled the editorial duties until 1719, at which time they passed to the physician James Jurin, who also succeeded Halley in the position of secretary in 1721.

What the first reviewer wrote prior to his list of five problems is a brief reference to De Moivre's new theory of recurring series. The motivation for the development of recurring series is related to the recursive nature in which the duration of play problem can be treated. At the conclusion of any game between the two players A and B, one of three things happens: A is ruined or exhausts his capital, and the series ends; B is ruined and the series ends; or play continues to the next game. In the current game, A is ruined if he has only one unit of capital left from the previous game and loses the current game. Consequently, A's probability of ruin at, say, the nth game is the probability that A loses the current game times the probability that the series has continued to the $(n - 1)$th game while leaving A with one unit of capital. A similar relationship holds for B. The sum of the two ruin probabilities for A and B at the nth game is the probability that the duration of play is exactly n games. The sum of the ruin probabilities for the first n games is the probability that the duration of play is at most n games.[9] This sum has the form of a recurring series.

Rather than launching into a discussion of recurring series at this point, let us focus instead on the fourth problem in the first reviewer's list dealing with the division of the circle. For the purpose of illustration, De Moivre assumed in his 1722 paper that players A and B each begin with capital of 10 units. He then takes a circle of radius 1 and divides the arc, defined by half the circumference of the circle, into 10 arcs of equal length. This is shown here with the diagram that accompanies De Moivre's 1722 paper. In the diagram, the lengths of the five (= 10/2) lines QF, OE, MD, KC, and HB are $\sin(\pi/10)$, $\sin(3\pi/10)$, $\sin(5\pi/10)$, $\sin(7\pi/10)$, and $\sin(9\pi/10)$, respectively. The numerical solution to the duration of play problem in this case is a function of the values of $\cos(\pi/10)$, $\cos(3\pi/10)$, $\cos(5\pi/10)$, $\cos(7\pi/10)$, and $\cos(9\pi/10)$.[10] This is reminiscent of De Moivre's work in 1707 in which he found the roots of a polynomial by dividing an arc of a circle into equal parts that correspond to the degree of the polynomial, an odd number. De Moivre's semicircle shows up in an allegorical engraving in *Doctrine of Chances*.

In order to understand De Moivre's allegorical engraving and what it says about Montmort,[11] it is necessary to understand another allegorical engraving, this one appearing in both editions of Montmort's *Essay d'analyse*. In the preface to the *Essay d'analyse*, Montmort states that one of his purposes in writing the book is to combat superstition. Games of chance are not subject to fickle fortune, but instead the outcomes follow mathematical laws which he demonstrated throughout the book. This idea is illustrated in the engraving shown here that appears at the head of the main body of the book in the second edition. Montmort commissioned Sébas-

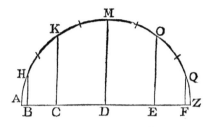

De Moivre's semicircle.

tien Le Clerc, one of France's leading engravers, to execute the work. Le Clerc was skilled in mathematics and had illustrated many scientific works.

The general scene shows a number of people playing at various games in a large room, probably at one in the royal chateaux of Louis XIV. The message is hidden in what is happening in the foreground of the scene. On the left is Montmort with Minerva, the goddess of wisdom. Minerva is identified by her helmet and the classical robes she is wearing. Montmort is identified by the monkey at the chessboard. The monkey's left hand is pointing to a bishop's initial position on the board. The position of the right hand shows a knight's move. Montmort, as Pierre Rémond, had been a canon in the Catholic Church, then left his position to get married and purchased an estate to become Sieur de Montmort—an initial canon's position followed by a knight's move into the laity. On the table where Minerva is seated are mathematical instruments, showing her wisdom in mathematics. Despite

The engravings at the beginning of Montmort's *Essay d'analyse.*

the goddess's wisdom in the subject, Montmort stands in a dominant position over her and she is copying what Montmort has written on the paper he is holding for her to see. On the paper are some mathematical jottings that symbolize Montmort's mathematical discoveries related to chance. On the right in the foreground the god Mercury is seated at a table with a man and a woman. The dice have just been thrown and the man has won. Mercury, the giver of good luck and presider over dice games, is taking credit for the man's good fortune. His perceived hold over the outcomes of dice games is soon to end because of Montmort's new work. Mercury is seated at the table rather than standing in a dominant position over the players. Similar to what is written in the preface that follows the engraving, the message is that chance outcomes are subject to mathematical laws, not to the whims of the gods. From Montmort's position as he appears in the engraving, the additional message in the engraving is that Montmort is responsible for the discovery of these mathematical laws.

As *Doctrine of Chances* was going to press, De Moivre wanted an engraving of his own to grace his new book. With the help of Brook Taylor, De Moivre commissioned the painter and etcher Joseph Goupy to design the engraving. Taylor knew Goupy well; he had studied drawing with him.[12] Once designed, the actual work of engraving was carried out by Bernard Baron, a young, up and coming engraver recently arrived in London from Paris. Taylor was involved in the design process with Goupy. Probably near the middle of September 1717, De Moivre received a print from Taylor that might serve as a frontispiece for the book. De Moivre replied to Taylor on September 28, thanking him for the print and expressing his pleasure in it.[13] Then he laid out what he thought should be in the engraving: Minerva and Mercury should be present with perhaps Minerva showing Mercury a piece of paper with something written on it related to probability theory, or the paper could be placed on a table with several people standing around it, and with one of the people pointing to it and instructing the others in probability theory. On the table should be a dice-box and dice "to signifie that ye truth of the Calculation may be confirmed by experiment." De Moivre suggested that any or all of three things could be on the paper. The first two were simple mathematical formulae; one was an expression related to his Poisson approximation to the binomial and the other was an infinite series expressing the solution to the problem of the pool when there are four players in the pool. These were associated with probability problems to which De Moivre could definitely claim originality, both in the statement and in the solution. And they were problems in which he took some pride. The third thing came with a detailed explanation. It was a

semicircle divided into seven or more equal parts with perpendiculars from every point of Division upon ye Diameter alternately markt with black and prick't lines thus this figure alluding to that part of my Book wherein I shew how ye Problem of ye duration of Play may be solved by ye versed sines of those arcs, in some cases taking ye versed sines which correspond to ye Black lines, in others taking ye versed sines which correspond to ye prickt ones, which discovery I take to be superiour to any thing which is in my Book, I conceal my Demonstration, and tho there are a great many things in my Book which may lead to it, yet I fancy that Mr de Montmort and his friend at Paris [Waldegrave] as well as young Bernoully will not easily find it.

Minerva, but not Mercury, made it into the final version of the engraving. The semicircle, but not the formulae, did as well. The table surrounded by people, with the dice and dice-box on the table, were also featured. The eventual form of De Moivre's allegorical engraving shows Minerva, on the left of the picture, pointing to a piece of paper with De Moivre's semicircle on it.[14] The paper actually shows the whole circle, which Matthew Maty, De Moivre's friend and first biographer, interpreted as De Moivre's diagram superimposed on a wheel of fortune.[15] The piece of paper is held by Fortuna, the goddess of fortune. She is identified by the wheel of fortune behind her and the cornucopia at her feet. With Minerva standing in a dominant position over Fortuna, the interpretation is that De Moivre's mathematical results dominate fickle fortune or fate. The paper under the cornucopia has some illegible writing on it. It may represent some previous work that has borne fruit, perhaps referring to Huygens's original results in *De ratiociniis*. The dice and dice-box are on a table as De Moivre requested. Added to the scene are four men standing around the table. The clean-shaven one is De Moivre. Comparing this engraving to Montmort's, De Moivre has replaced Mercury at the gaming table and is instructing the three other men on the theory of probability.

The scene on the right of the engraving is a gathering of mortals. With a subtle swipe at Montmort, De Moivre expresses through the engraving that he does not have the effrontery to speak directly to the gods and instruct them. The other men at the table, all dressed in classical robes, are ancient philosophers and mathematicians receiving instruction from the modern mathematician. Eighteen years after the publication of *Doctrine of Chances*, in a book review for the Royal Society[16] of Martin Kahle's *Elementa Logicae Probabilium*,[17] De Moivre made explicit the connection between the ancient and eighteenth century approaches to probability:

This doctrine was first introduced by Huygens in the year 1657, in a little Treatise intitled: Ratiocinia de Ludo Aleae: and has since been followed by James Bernoulli, Monmort, Nicholas Bernoulli, myself and perhaps others. For altho' the ancients had the same Idea of Probability as we have; and they mentioned several degrees of it, as appears in the words *Probabile* and Probability often

The engravings at the beginning of De Moivre's *Doctrine of Chances*.

used by Cicero & other writers, yet the Author [Kahle] observes, that the distinct measure of it was never assigned till the times above mentioned.

Clearly, in the engraving De Moivre is bringing the ancients up-to-date in modern probability.

The middle part of the engraving has additional swipes at Montmort. Two naked boys are sitting with a pair of dice lying at their feet. A short distance away are some discarded cards and further yet is a chessboard of size 4 × 6 squares rather than the standard 8 × 8 as shown in Montmort's engraving. One of the boys is reading a book, perhaps *Doctrine of Chances*, to the other explaining De Moivre's newly discovered results in probability. The discarded chessboard, being incomplete, is an indication that the work in the *Essay d'analyse* is also incomplete.

In the text of *Doctrine of Chances*, De Moivre strictly adheres to the rules of the Republic of Letters. He is polite to Montmort and praises his work in various places. The engraving provides no actual words that are critical of Montmort, but the interpretation of the figures in it provide a direct response to Montmort's scathing attack on De Moivre's work that appeared in *Essay d'analyse*. Montmort's attack broke the rules of the Republic of Letters; De Moivre's response did not. In words, all is politeness from De Moivre. In pictures, De Moivre essentially says that his work in probability is better than Montmort's. He had also done a better job at banishing Fortuna from games of chance than Montmort had.

For some time, as early as 1716, Montmort wanted to see a copy of De Moivre's book. As time went on, he became impatient with the delay. He wrote to Brook Taylor on December 17, 1717, expressing his impatience and wrote again on January 25, 1718, with a similar sentiment. In the latter letter, he asked Taylor to send him a copy as soon as it was in print. He reiterated this desire in a letter of February 29, 1718.[18] By June 1718, Montmort had a copy of the book. He wrote to Johann Bernoulli that De Moivre had lifted material from both editions of *Essay d'analyse* including everything Montmort had written on combinations. He told Bernoulli that it had angered him for about a day.[19] Bernoulli was sympathetic. He was still smarting from the accusations made against him by John Keill and believed that other Englishmen had plagiarized his work. Now seeing De Moivre as completely anglicized, since De Moivre had written his book in English, he expressed to Montmort the opinion that it was now natural for De Moivre to plagiarize as all the English did it, while at the same time crying thief.[20]

Over the next few months Montmort read *Doctrine of Chances* very carefully. In February 1719, he sent an eight-page letter to Brook Taylor with a list of thirty-three criticisms.[21] He also sent a copy of the letter to Johann Bernoulli.[22] Sprinkled throughout the list are comments that led Montmort to the same charges as before with *De Mensura Sortis*: there was nothing new in much of the book; Montmort had already obtained many of the results; and De Moivre had used Montmort's results without crediting him. Again, Bernoulli was sympathetic but could not comment on how bad the plagiarism was since he had not read the book.

After reading Montmort's lengthy critique, Brook Taylor had a few comments of his own. He had read De Moivre's book. It is uncertain if his comments were ever sent to Montmort; they may have been sent to De Moivre. In St. John's College Library where Taylor's correspondence is kept, Montmort's criticisms of De Moivre are placed in Montmort's correspondence with Brook Taylor, while Taylor's comments are placed in De Moivre's correspondence with Taylor. One of Taylor's comments goes directly to the plagiarism issue and, at the same time, touches briefly on what might or might not be new in *Doctrine of Chances*. Taylor writes:

> It is very true that most of the things in Mr Moivres Book are to be found (at least with some little difference, either in the manner of stating the Problems, or in the Method of Solution,) in Mr Montmorts. This would have been an unpardonable Theft, if Mr Moivre had industriously avoided mentioning of Mr Montmorts Book, or had endeavour'd to conceal his own having read it: but as he had done the contrary, has own'd himself to have read Mr Montmorts Book with care, has recommended it with encomiums, has particularly taken pains to set forth the excellency of some places he thought most remarkable in it; this ought rather to be esteem'd as a mark of the value that Mr Moivre had for that Book, and of the respect he had for its Author that he would so far copy it, than as any disrespect, or unfair dealing. Whoever reads Mr Moivres Book will naturally be directed and be made desirous to read Mr Montmorts, and there he will see how very much it is probable Mr Moivre may have learnt from him.[23]

De Moivre's trigonometric solution to the problem of the duration of play was certainly new. The question then is, What else was new in *Doctrine of Chances*? And a related question is, How much further did De Moivre take the mathematics of chance beyond what he had already done in *De Mensura Sortis* and what Montmort and Bernoulli had obtained in the second edition of *Essay d'analyse*?

In terms of the old material in *De Mensura Sortis*, with one exception De Moivre reproduced all the results in his *Doctrine of Chances*. Often the Latin statement of the problem from *De Mensura Sortis* as rendered in *Doctrine of Chances* is slightly different, and clearer, than a word-by-word translation from Latin to English. Sometimes De Moivre shortened or simplified the proof to a problem in *De Mensura Sortis*, especially the problems related to the duration of play. In other problems there is an expansion of material; more examples are given and more remarks are made about the problem in question. The big change to the *De Mensura Sortis* problems is in Robartes's lawn bowling problems. Montmort had written to Nicolaus Bernoulli on September 5, 1712, saying that De Moivre had only given solutions to simple cases of a more general problem and not a general solution,[24] even though De Moivre stated in corollaries to his problems that the general solution followed along the lines of argument that he used. Montmort provided such a general solution in the second edition of the *Essay d'analyse*—an arbitrary number of points left to win and differing skills between the players.[25] De Moivre's response in *Doctrine of Chances* was to provide his own general solution that he claimed was simpler to obtain than Montmort's. After all his complaints about De Moivre not giving him any credit, it is interesting to note that when Montmort states the general problem and solution in the main body of *Essay d'analyse*, there is no mention of De Moivre.

Some parts of *Doctrine of Chances* that irritated Montmort should not have if De Moivre had provided a little more background to his mathematical approaches to problems. When Johann Bernoulli wrote Pierre Varignon in 1712 that he found De Moivre's way of solving probability problems a little obscure; the same observation might be applied to some of De Moivre's more general comments on the problems he solved. Problem 11 in both *De Mensura Sortis* and *Doctrine of Chances* provides an illustration. It is a solution to one of Huygens's challenge problems from *De ratiociniis*. The problem deals with three players in turn drawing black or white balls from an urn without replacement until a white ball is drawn. Originally, Montmort had dealt with the problem when the balls are drawn with replacement. In *Doctrine of Chances* De Moivre adds a new solution to the without-replacement problem.[26] First he shows that the solution is the sum of a finite sequence of numbers, say $a_1 + a_2 + \ldots + a_n$. If n is large, then it may be onerous to evaluate the sum directly. To illustrate the difficulty, De Moivre gives an example of one hundred balls of which four are white. The summed sequence of $n = 33$ terms needed to evaluate the chances for the first player is $1 \cdot 2 \cdot 3 + 4 \cdot 5 \cdot 6 + 7 \cdot 8 \cdot 9 + \ldots + 97 \cdot 98 \cdot 99$. De Moivre finds the sum using finite differences, obtaining

$$\sum_{i=1}^{n} a_i = a_1 + \frac{m}{1}a_2 + \frac{m(m-1)}{1\cdot 2}\Delta a_2 + \frac{m(m-1)(m-2)}{1\cdot 2\cdot 3}\Delta^2 a_2 + \frac{m(m-1)(m-2)(m-3)}{1\cdot 2\cdot 3\cdot 4}\Delta^3 a_2 + \cdots,$$

where $m = n - 1$, $\Delta a_2 = a_3 - a_2$, $\Delta^2 a_2 = \Delta(a_3 - a_2) = a_4 - 2a_3 + a_2$, and so on. The terms in the sum on the right of the equation continue until they become 0. In the numerical sequence that has been given, it may be verified that $\Delta^4 a_2 = 0$, as well as all subsequent terms.[27] De Moivre wrote that his formula was obtained from some of Newton's results on finite differences. Initially, he mentions *Principia Mathematica* and then cites *Methodus Differentialis*, which Newton had written years before but had not published until William Jones put out his collection of Newton's papers in 1711.[28] This incensed Montmort. He had obtained a formula for the solution of the without-replacement problem in his second edition of *Essay d'analyse* and De Moivre had ignored it.[29] Why cite a result that leads to the solution rather than the solution itself, Montmort wanted to know. That kind of behavior violated the code of the Republic of Letters, according to Montmort. What Montmort did not know, and what De Moivre never stated, was that he had previously obtained his formula in 1708, well before Montmort found his, and had sent it in a letter to Johann Bernoulli. By stroking Newton's ego, he had rubbed Montmort the wrong way.

One basic difference between *Essay d'analyse* and *Doctrine of Chances*, as well as earlier in *De Mensura Sortis*, is in the two mathematicians' approaches to probability. Although he developed a number of results in combinations that could be used to enumerate favorable and unfavorable events, Montmort continued to adhere to Huygens's approach to probability through expectations. De Moivre, on the other hand, set out his definition of probability very clearly at the beginning of *Doctrine of Chances*, followed by the description of an approach that has come to be known as the classical theory of probability. In De Moivre's words,

> The Probability of an Event is greater, or less, according to the number of Chances by which it may Happen, compar'd with the number of all the Chances, by which it may either Happen or Fail.[30]

Relying on an eighteenth-century use of the word *doctrine*, the doctrine of chances is then the system of principles that leads to the valuation, exact or approximate, of the number of chances that can be attributed to an event happening or failing. Following his initial definition of probability, De Moivre then went on to define odds and expectations, as well as dependent and independent events.

One difference between *De Mensura Sortis* and *Doctrine of Chances* is in the treatment of some actual games of chance that were played at the time the book was written. In *De Mensura Sortis* there are only hints at two of these games. The problem of the pool, or Waldegrave's problem, derives from a way to increase the number of players in the card game *Piquet*, a game never explicitly mentioned in *De Mensura Sortis*. Robartes's lawn bowling problems were more a way to complicate the division of stakes problem than an actual analysis of the game. By comparison, in *Doctrine of Chances* De Moivre analyzes the card games *Basset* (or *Bassette* as De Moivre spelled it), *Faro* (or *Pharaon*), *Piquet*, and *Whist* (or *Whisk*), and the dice games *Hazard* and *Raffle* (or *Raffling*), as well as some British-style lotteries. There is an enormous overlap with the *Essay d'analyse*. Montmort dealt with all the same card and dice games with the exception of *Whist*. *Whist*, the exception, is related to the card game *Ombre*, which Montmort did analyze. Montmort also analyzed a French lottery run in a style similar to the British ones in which winning tickets are drawn from the complete set of tickets rather than winning numbers drawn as in a number lottery or lotto.

With all the games he analyzes, it is difficult to say if De Moivre ever played any of them. If one had to choose a most likely candidate, it would be *Whist* since it was played regularly at Slaughter's Coffeehouse, which he frequented. Another likely candidate is *Piquet* since De Moivre picks the game to illustrate some points when discussing the philosophy of chance in the preface to the book. He could have picked several other games for his exemplar.

In some of the games, De Moivre provides essentially the same analysis as Montmort. In others, he goes beyond what Montmort had written. For example, while Montmort analyzed a few cases in the dice game *Hazard*, De Moivre deals with all the possible situations in the game.[31] Montmort's response to this difference is that he had the essentials and so De Moivre should have recognized Montmort's priority by citing his results.[32]

What was De Moivre's motivation in spending energy and space re-examining what Montmort had done? This baffled Montmort. All that he could see was De Moivre copying his results from *Essay d'analyse* without crediting him, and that angered him. There is a hint in the text for one of De Moivre's motivations for working on some of these games. *Whist*, *Piquet*, *Hazard*, and *Raffles* appear at the end of *Doctrine of Chances*, almost as a postscript. De Moivre used these games as useful exercises to illustrate the use of combinatorial mathematics. Perhaps his work as a tutor was showing through—teach the theory and then practice the concepts with examples familiar to the reader. Other motivating factors may be found by examining some of the games in a little more detail.

Consider the dice game *Raffle*.[33] Five and ten years before, in both editions of *Essay d'analyse*, Montmort described the game and provided a table containing the chances of winning at each of the possible throws of the dice. In his treatment of the problem, De Moivre does not bother with the description. Instead, without giving any rules or analysis of the game, he launches directly into a description of a table from which the chances of winning can be obtained for the various situations in the game. Montmort's and De Moivre's tables are lain out differently, but correspond almost exactly when it comes to the enumeration of the chances to win. The "almost" refers to one number. Montmort has 38,867 in one spot while De Moivre has 33,867, leaving one with the impression that De Moivre had recalculated or checked all of Montmort's calculations very carefully. Such extreme overlap did not go unnoticed by Montmort. After noting the similarities in the tables, Montmort thought that perhaps the table had been published because De Moivre had found, but had not explicitly commented on, the typographical error in Montmort's table. Montmort's 38,867 should have been the number that De Moivre gave.[34] Montmort had obviously read De Moivre's book very closely. De Moivre's real motivation only became apparent twelve years later when he published his *Miscellanea Analytica*. There he states that his table had been constructed by Francis Robartes during the 1680s or 1690s.[35] De Moivre had probably never played the game. His interest in the problem is more likely seen in the final few sentences of his treatment of *Raffle* where he used Robartes's table and a binomial expansion to find the probability that a given player will win when there is a specified number of players in the game.

While the other games may have been afterthoughts, De Moivre treats *Basset* and *Faro*, which are variations on the same type of card game, near the beginning of *Doctrine of Chances*.[36] Where De Moivre has placed the analysis of these games in the text is important. The analyses of *Basset* and *Faro* appear immediately after Problem 11, De Moivre's solution to one of Huygens's challenge problems. On one level, De Moivre has given early on in his book an analysis of a game, *Basset*, which was very familiar to the subscribers of his book. On another level, De Moivre has provided an application for the result he obtained for Huygens's challenge problem, the statement of which he had originally interpreted differently from Montmort. The solutions to probability problems in *Basset* and *Faro* rely on the solution to this challenge problem. Just as the challenge problem treats the selection of white and black balls without replacement, in both card games a succession of pairs of cards are drawn by the banker, or house, from a deck without replacement and compared to a card held by the person playing against the bank.

Contrary to his treatment of the games at the end of *Doctrine of Chances*, De Moivre laid out the rules for both *Basset* and *Faro*, just as Montmort did in *Essay d'analyse*. The rules for *Faro* were the same in England and France, while the rules for *Basset* were slightly different. Just as Montmort did, De Moivre produced a table each for *Basset* and *Faro* that shows the advantage that the banker holds at each step of the game, a step being the draw of a pair of cards from the deck. De Moivre's table for *Faro* is the same as Montmort's since the rules of the game were the same in each country.

Basset had been analyzed previously by Joseph Saveur in 1679 and by Jacob Bernoulli in his *Ars Conjectandi* published in 1713.[37] Montmort's analysis of *Basset* is relatively complete. Consequently, he felt that his work on these card games should have been cited by De Moivre, and he considered it dishonest of De Moivre not to do so as he was treating these games.[38] In his preface to *Doctrine of Chances*, De Moivre had only mentioned in his typically vague way that "several great Mathematicians" had found the banker's advantage at the various stages of the game.

De Moivre was not plagiarizing or blindly copying Montmort's work. Beyond being a showpiece for his approach to the solution to Huygens's challenge problem and providing some satisfaction to some of his aristocratic subscribers, another motivating factor was that De Moivre wanted to come up with a single measure for the advantage held by the banker in these games rather than measures at each stage of the games. This was new. He was able to find such a measure, but it depends on the restrictive assumption that the card held by the person playing against the bank is selected at random. In normal play of the game, the selection is non-random, so the player can adopt some strategies of play as the game progresses.[39] Random selection runs counter to gambling intuition.

Did De Moivre ever play *Basset* or *Faro*? The quick answer is probably not, or at most rarely. On the other hand, it is very likely that he was very familiar with *Basset* but not *Faro*; *Faro* did not become popular in England until later in the eighteenth century. The "probably not" answer comes from the fact that *Basset* was notorious for its ability to ruin players, and De Moivre seems to have been a person who was very careful with his money. He arrived a refugee with probably a small amount of money, and his friend and first biographer stated: "Mathematics did not make him rich and he lived a mediocre [or modest] life bequeathing his few possessions to his next-of-kin."[40] De Moivre left this world holding £1600 capital in annuities, a very tidy sum for the time.

Basset was all around De Moivre: with the families whose children he taught, with his Huguenot friends, and at the London stage that he sometimes frequented, a short walk away from his lodgings in St. Martin's Lane.[41] At a minimum, he would have been familiar with the game, even if he did not play it. *Basset*, because of the high stakes involved, was played by the nobility and other landed interests[42] who supplied De Moivre with their sons to teach. A few lines from the 1701 play *Sir Henry Wildair* by George Farquhar connect the nobility, the game, and the stage. Addressing the heroine of the play, Lady Lurewell, Sir Henry says,[43]

> What, forswear Cards! Why, you'll ruin our Trade.—I'll maintain, that the money at Court circulates more by the Basset-Bank than the Wealth of Merchants by the Bank of the City. Cards! the great Ministers of Fortune's Power; that blindly shuffle out her thoughtless Favours, and make a Knave more pow'rful than a King.

Basset is mentioned in several plays on the London stage at the time, including one by Susanna Centlivre called *The Basset Table*, first performed in 1705.[44] These plays were written, in part, as a response to the clergyman Jeremy Collier's influential call in 1698 for more morality in the theater. Both *The Basset Table* and *Sir Henry Wildair* were performed at Theatre Royal in Drury Lane, just over a mile walk from the upper end of St. Martin's Lane. Another play that hit the stage of the Theatre Royal, this one in 1699, was Abel Boyer's English adaptation of Jean Racine's tragedy

Iphigénie.[45] The epilogue to the play was written by Pierre-Antoine Motteux. Like Motteux and De Moivre, Boyer was another Huguenot who was part of the circle of émigré intellectuals that met at the Rainbow Coffeehouse.[46] In the published version of the play, both Boyer and Motteux make reference to *Basset* as a drain on the purse, Boyer in his epistle dedicatory and Motteux in his epilogue.

Whether De Moivre and his friends ever discussed the evils or pitfalls of playing *Basset* over their drinks at the Rainbow Coffeehouse will never be known. What is known, from Abbé Jean-Bernard Le Blanc's visit to De Moivre in the late 1730s is that Le Blanc "did not perceive the he [De Moivre] had ever calculated the effects of gaming, with regard to morality, though that is a much more essential thing than the theory of chances."

Although his Huguenot friends may have had some influence on his thoughts about particular games of chance, they are not mentioned in *Doctrine of Chances*. On the other hand, his closest English friends are mentioned in the text: Newton, Robartes, Jones, Taylor, and Halley. The book is dedicated to Newton. One Scotsman is alluded to, but never named outright in the book. At the beginning of the preface, when De Moivre is writing of the time when he read Huygens's *De ratiociniis*, he mentions "a little English Piece" written by "a very ingenious Gentleman." This is polite, but not overly friendly, as when he writes of William Jones as his "intimate friend" or Halley as his "respected friend." The ingenious gentleman is John Arbuthnot who translated Huygens's book and augmented it with results of his own in 1692.[47] Perhaps De Moivre still bore some small general resentment of the Scots related to his little dustup with George Cheyne that was now several years old.

There is one more Englishman mentioned in the text, an obscure gentleman named Thomas Woodcock who suggested a problem for De Moivre to work on.[48] The problem was new, so Montmort had no comment on De Moivre's solution. Woodcock was not a fellow of the Royal Society. But he had money and position; he married a widowed aunt of Thomas Pelham-Holles, later 1st Duke of Newcastle-upon-Tyne.[49] De Moivre probably tutored Pelham-Holles in his youth.

The problem that Woodcock suggested is a generalization of the gambler's ruin problem, which was originally Huygens's fifth challenge problem that De Moivre had solved with an ingenious mathematical trick in *De Mensura Sortis*. The gambler's ruin has been described already in the following terms: Two players, call them A and B, engage in a series of games where A has probability p of winning any game and B has probability q. The winner of any game is given one unit from the loser's capital and the series ends when one of the players is ruined or his capital is reduced to zero. The two players may start with different amounts of capital. Woodcock's twist on the problem is to separate the amount "units of capital" into two parts: chips (or *stakes*, in De Moivre's terminology) to be won and the amount bet by the two players. In Woodcock's scenario, A has a chips to play with and B has b chips. Every time a game is played, A puts down an amount α and B puts down an amount β so that the pot to be won grows by $\alpha + \beta$ at each game. As the series progresses chips are won and lost by each player, and the game ends when one of the players has accumulated $a + b$ chips and hence has obtained the entire capital put in by the players. What De Moivre calculated is the advantage or disadvantage that A has over B.

Like De Moivre's attempt to find the banker's overall advantage in *Basset*, this probability problem seems to have no basis in gambling reality. Suppose that A's total monetary resources amount to an. If the series of games goes beyond n in number, then A cannot add money to the pot. De Moivre's solution to Woodcock's problem might be described as mathematically elegant, but socially irrelevant. Once he solved Woodcock's problem, De Moivre carried on further down the mathematician's rather than gambler's path and went on to make a generalization to Woodcock's generalization. He assumed that the amount put in by each player increases in arithmetic progression as the games continue. At game n in the series, A puts down an amount an and B puts down βn.

De Moivre sent Woodcock's problem to Nicolaus Bernoulli with a hint of how he had solved it. The hint was that he had used an infinite series approach. Bernoulli ran with the hint and sent back two solutions to De Moivre. One is very short and uses infinite series methods that De Moivre used throughout the *Doctrine of Chances*, and so De Moivre included it as a postscript to his own solution. Since he had stopped writing to either Johann or Nicolaus Bernoulli by the middle of 1714, De Moivre must have obtained his own solution and sent Bernoulli the problem before that time.

There is some internal evidence that the manuscript version of De Moivre's solution to Woodcock's problem that became the printed text did not change between 1714 and 1718. In a numerical example, De Moivre takes α as one guinea and β as twenty shillings. The numerical solution only works when the guinea is worth 21.5 shillings. This was the value of the guinea from 1699 to 1717; as Master of the Mint, it was Newton who recommended the change in value to the standard 21 shillings in 1717.[50] It is likely then that what was given to the printer to typeset for *Doctrine of Chances* was not a single seamless manuscript from beginning to end, but a number of manuscripts pieced together from De Moivre's new results along with the translation of, and amendments to, the problems in *De Mensura Sortis*.

Further evidence of the patchwork nature of parts of *Doctrine of Chances* is De Moivre's treatment of what is known as the theory of coincidences and the problem immediately preceding this theory. The theory of coincidences is handled in Problems 25 and 26 in *Doctrine of Chances*. In two places in the preface, De Moivre states that the theory appears in Problems 24 and 25. Problem 24, as it appears in *Doctrine of Chances*, is an exercise in the application of the formula for the sum of an infinite geometric progression. Problem 24 may have been inserted at a late date as a replacement for the one problem in *De Mensura Sortis* (Problem 13 there) that did not make it into *Doctrine of Chances*. Both problems involve two players, A and B, engaging in a series of "throws" of a randomizing device such that A's chance of winning a throw is p and B's is q. In the problem given in *Doctrine of Chances*, $p + q = 1$ while in *De Mensura Sortis* $p + q < 1$. In *Doctrine of Chances*, A and B each bet one unit at each throw in the series and the throws continue until the first time that B wins. The object is to find the expected gain for player A. In *De Mensura Sortis*, one unit is staked by A and B at the beginning of the series. Player A throws once, then B and A successively throw twice until one of them wins a throw and hence the stake. The object is to find the probabilities of winning for each player.

A coincidence in mathematical terms is a type of match. For example, take two decks of cards, shuffle them, and start dealing face up from each deck. If the two most recently turned cards are the same, then there is a match or coincidence. More generally, one can think of a list of objects that has been randomized. A coincidence happens when a randomized object shows up in the same position that it was on the original list. For Montmort, Nicolaus Bernoulli, and De Moivre, the problem of coincidences came in two forms: when the objects on the list are all distinct, such as cards in a deck, and when the objects are comprised of different groups of similar objects, such as the denomination of cards ace through king without regard to the suit.

De Moivre's solution to the theory of coincidences shows another side of De Moivre as a mathematician. Earlier in *De Mensura Sortis*, we saw him using the approach of simplification followed by generalization when solving the problem of the pool for three players. This is a typical approach when solving a practical problem that has some complexities to it, in this case the pool as actually played by gamblers. In the theory of coincidences, De Moivre searched for an elegant solution and found it. In mathematical terms, an elegant solution is one that is short and can be easily generalized to a whole family of similar problems. It can also be based on new and original insights.[51] De Moivre's solution is short. It is also very general in that it includes all the cases considered by Montmort and Nicolaus Bernoulli. To put icing on the cake of elegance, De Moivre developed what he called a "new sort of Algebra" to obtain his results.

In modern terminology, what De Moivre essentially developed was a new method for finding probabilities of various compound events when the events are exchangeable. Suppose we have n events from which a number k of them is considered. By exchangeable, it is meant that the probability of the joint occurrence of these k events is the same as the probability of the joint occurrence of any other k events out of the n. The compound events that De Moivre considers are that out of n objects (or events) the first m are matches (or the event occurs) and the remaining $n - m$ are not matches (or the event does not occur). Since the events are exchangeable, it does not matter the order in which the matches and mismatches occur. Consequently, the probability of any m matches and any $n - m$ mismatches is the probability that the first m match and the rest do not, multiplied by the number of ways that the matches and mismatches can be permuted. De Moivre obtains a simple expression for the probability that the first m match and the rest do not.

In modern notation, he obtains

$$P\left(O_1 O_2 \cdots O_m \bar{O}_{m+1} \cdots \bar{O}_n\right) = \sum_{i=0}^{n-m} (-1)^i \binom{n-m}{i} P(O_1 \cdots O_{m+i})$$

where O_i is the event that ith object is a match and $\bar{C}_i$ is a mismatch.[52] It is a very easy algorithm to describe in words, as De Moivre did. For m matches and $n - m$ mismatches, write down the probabilities of the first m matching without regard to the remaining objects, then the first $m + 1$ matching, again without regard to the remaining objects, and so on. Attach alternating signs to this list of probabilities and make the coefficients of the probabilities the set of binomial coefficients obtained from the expansion of $(1 + 1)^{n-m}$. Here, and elsewhere in his *Doctrine of Chances*, De Moivre was carrying over from *De Mensura Sortis* his theme of the use of binomial expansions to solve probability problems. Unable to see beyond De Moivre's lack of citation of his own work on coincidences, Montmort could not appreciate the elegance of De Moivre's solution.[53] On the other hand, when Nicolaus Bernoulli presented De Moivre with an elegant solution to Woodcock's problem, De Moivre acknowledged it and included it in his book, even though he had stopped communicating with Bernoulli.

Woodcock's problem involves an infinite series solution, as does the solution to the problem of the pool or Waldegrave's problem. In *De Mensura Sortis*, De Moivre had solved the problem for three players but said his method could be generalized to more than three. As mentioned in Chapter 7, De Moivre had obtained the generalization to four players in 1714 and published the solution in *Philosophical Transactions*. Since the article was in Latin, it was aimed at an international audience. What appears in *Doctrine of Chances* for the most part is a very faithful English translation of the *Philosophical Transactions* article.[54] De Moivre calls the four players A, B, C, and D. He assumes that A and B play first, and that B wins the first game. For A to win the pool he must repeatedly come back into the game until he has won three in a row, one against each of B, C, and D, while the other three cannot have won three in a row. Since there are four players and a player must win three in a row to win the pool, the pool can only be won on the third game or afterward. De Moivre denotes the probability that A wins on game numbers 3, 4, 5, 6, 7, ... by A', A'', A''', A'''', A^V, ..., respectively. Given that A has lost the first game and cannot come back until the fourth game, and so cannot possibly win the pool until game number 6, then $A' = A'' = A'''= 0$. De Moivre has a similar notation for B, C, and D winning the pool on games 3, 4, 5, etc. There are recursive relationships between the probabilities. For example, the recursion relating A's probabilities to C's and D's at game 6 is given by

$$A'''' = \frac{1}{2}D''' + \frac{1}{4}C''$$

in De Moivre's notation. With the appropriate changes to the superscripts, this recursive relation holds for any game number. The probability that A wins the pool is the infinite sum

$$A' + A'' + A''' + A''' + A^V + \&c.$$

Despite many who would agree with Montmort in this case that De Moivre's solution is "cumbersome," including Anders Hald who has put De Moivre's solution into a modern format, De Moivre was very proud of it. In his discussions with Brook Taylor about what should be included in the allegorical frontispiece to *Doctrine of Chances*, De Moivre suggested that the expression $A' + A'' + A''' + \&c.$, along with perhaps another one related to the Poisson approximation to the binomial, should appear on a piece of paper lying on a table. Although this did not happen, Taylor did have some further involvement in the problem of the pool. He gave De Moivre a simple way to obtain the recursion relationships. Taylor's method is reproduced in *Doctrine of Chances* at the end of the treatment of the problem of the pool for four players.

Where De Moivre shines in the use of infinite series is in his development of the theory of recurring series[55] in which he applies the sum of the series to various aspects of the duration of play problem. As in *De Mensura Sortis*, the duration of play problem is the grand finale to *Doctrine of Chances*; the treatment of various games of chance at the end is like an encore. A recurring series is defined through the power series whose sum is

$$S = c_0 + c_1 x + c_2 x^2 + c_3 x^3 + \cdots$$

where the coefficients c_0, c_1, c_2, c_3, etc., have a specific relationship to one another; specifically, the current coefficient is defined in terms of several preceding ones. The number of preceding coefficients is called the order of the series. For example, in a recurring series of order two, a coefficient depends on the two immediately previous coefficients. For the ith coefficient, $c_i = a_1 c_{i-1} - a_2 c_{i-2}$, where a_1 and a_2 are known constants and the initial coefficients c_0 and c_1 are given as well. Higher order recurring series are similarly defined.

De Moivre's proof for the sum S of a general recurring series is a typical eighteenth-century style of proof. Prove the theorem for recurring series of order two, then order three, and then order four. Note the general pattern and write down the general result. As De Moivre expresses this last step, "The Law of continuation of the Theorems being manifest, they may all easily be comprehended under one general Rule." Modern mathematicians might decry the lack of rigor to the proof, but it did work for De Moivre and many others.

De Moivre's proof for an order two recurring series is very simple. He considers each term in the series. The ith term is given by $c_i x^i$, which can be denoted by t_i so that

$$S = c_0 + c_1 x + c_2 x^2 + c_3 x^3 + \cdots = t_0 + t_1 + t_2 + t_3 + \cdots.$$

From the recurrence relationship, $c_i = a_1 c_{i-1} - a_2 c_{i-2}$, we have $c_i x^i = a_1 c_{i-1} x^i - a_2 c_{i-2} x^i$ or equivalently $t_i = a_1 t_{i-1} x - a_2 t_{i-2} x^2$. De Moivre set this all out in columns and added on the initial terms. This gives

$$t_0 = t_0$$

$$t_1 = t_1$$

$$t_2 = a_1 t_1 x - a_2 t_0 x^2$$

$$t_3 = a_1 t_2 x - a_2 t_1 x^2$$

$$t_4 = a_1 t_3 x - a_2 t_2 x^2$$

$$t_5 = a_1 t_4 x - a_2 t_3 x^2$$

$$\vdots$$

The sum of the column to the left of the equals sign is S. The sum of the first column to the right of the equals sign is $t_0 + t_1 + a_1 S x - a_1 t_0 x$ and the sum of the second column is $a_2 S x^2$. Consequently, $S = t_0 + t_1 + a_1 S x - a_1 t_0 x - a_2 S x^2$. Solving for S yields

$$S = \frac{t_0 + t_1 - a_1 t_0 x}{1 - a_1 x + a_2 x^2}.$$

A similar method works for higher-order recurring series. The number of columns to the right of the equals sign will be the order of the recurring series.

Immediately after De Moivre's treatment of recurring series, he gives another method of summing a power series, which he says was sent to him by Montmort. If the coefficients in the power series given by c_0, c_1, c_2, c_3, etc., are such that for some order of finite differencing d, $\Delta^d c_0 = 0$, then the sum of the series is given by

$$\frac{x}{1-x} \left\{ c_0 + \frac{x}{1-x} \Delta c_0 + \left(\frac{x}{1-x} \right)^2 \Delta^2 c_0 + \cdots + \left(\frac{x}{1-x} \right)^{d-1} \Delta^{d-1} c_0 \right\}.$$

The result is reminiscent of De Moivre's summation of a sequence by finite difference methods that he had obtained in 1708.

One would think that Montmort would finally be satisfied. He was cited and acknowledged. Instead, the tap of complaint was opened in full. In his letter to Brook Taylor, Montmort complains that De Moivre's work on recurring series is derivative of Nicolaus Bernoulli's work. Montmort quoted from a letter dated August 30, 1714, that he had received from Nicolaus Bernoulli in which Bernoulli said that he had solved a problem in infinite series which had a recurrence relation in it.[56] Montmort had informed De Moivre of the result in 1715, just before Montmort made his trip to England. Taylor's only response to this accusation was

> I know nothing of the letter mention'd in this Article. If Mr Moivre had it, I cant say but it might give him a hint for the series he here treats of, Mr Bernoulli's series being a particular case of Mr Moivres; and therefore Mr Moivres might be found possibly by extending Mr Bernoulli's to a greater degree of generality.[57]

The reviewer of the manuscript opened at the Royal Society meeting in 1720 considered De Moivre's work in recurring series to be new and original. However De Moivre came by the idea for his work in recurring series, history seems to have sided with the anonymous reviewer about who was the original developer of the theory.

The theory behind the semicircle and its associated trigonometric formula to solve the duration of play problem was only hinted at in *Doctrine of Chances*. Even after a reading of the paper that De Moivre wrote in 1720 and published in 1722 in *Philosophical Transactions*, the proof of the result is not at all clear, although De Moivre expanded on the derivation in *Miscellanea Analytica* in 1730. Echoing Johann Bernoulli's comment over two hundred and seventy-five years earlier that he found De Moivre's method of solving problems a little obscure, Anders Hald made a similar comment in that he found that "De Moivre's proof in *Miscellanea Analytica* is somewhat incomplete."[58] It was Ivo Schneider working in the 1960s who was able to provide a satisfactory reconstruction of De Moivre's proof.[59] Leaving the reader to examine the mathematical details in Hald's or Schneider's work, I will only point out the connection between recurring series and the trigonometric solution. De Moivre showed that the sum of a recurring series can be expressed as the ratio of two polynomials, where the order of the polynomial in the denominator is the same as the order of the recurring series. Suppose the polynomial in x in the denominator of the ratio is of degree d. This can be expressed algebraically as $b_0 + b_1 x + b_2 x^2 + \dots + b_d x^d$. De Moivre showed in his 1722 paper that the reciprocal of this polynomial can be written in an interesting way that involves the roots of the polynomial. He obtained

$$\frac{1}{b_0 + b_1 x + b_2 x^2 + \dots + b_d x^d} = \frac{f_1}{1 - r_1 x} + \frac{f_2}{1 - r_2 x} + \frac{f_3}{1 - r_3 x} + \dots + \frac{f_d}{1 - r_d x},$$

where $r_1, r_2, r_3, \dots, r_d$ are the roots of the polynomial equation $b_0 + b_1 x + \dots + b_d x^d = 0$ and $f_1, f_2, f_3, \dots, f_d$ are some specific functions of the roots. In his 1707 paper, with further explanation of it in another 1722 paper,[60] De Moivre used trigonometric methods to find the roots of a certain polynomial equations. And so we come full circle back to the semicircle.

❧ 9 ❧
Doctrinal Dissemination and Further Development

Publication of *Doctrine of Chances* confirmed De Moivre as one of the leading mathematicians of his time in Britain. Knowledge of his work went beyond his circle of friends, mathematicians, Royal Society members, and aristocratic employers. Still within the circle, but at the periphery, was the actor and playwright, Colley Cibber. De Moivre and Cibber had some common connections; Cibber's original patron was the 1st Duke of Devonshire and he had other prominent Whig connections that included Robert Walpole.[1] The one known interaction between De Moivre and Cibber is that they discussed the difference between French and English actors' methods of declamation in tragedies.[2] Whether or not he had ever read it, Cibber was aware in some way of the *Doctrine of Chances* and must have felt that many in his audience knew of the book as well as of the author's reputation. Cibber was also an inveterate gambler. There is a gambling reference to De Moivre in Cibber's play *The Provok'd Husband*, which was first staged in 1728. One of the characters in the play is a gambler named Count Basset (after the card game). During the second act, Basset mentions that the "Demoivre Baronet" had lost a lot of money the previous night at White's Coffeehouse, a favorite haunt of gamblers.[3] Perhaps De Moivre was in the audience on opening night to hear his name.

De Moivre's reputation in the mathematics of gambling went beyond even his wider circle of acquaintances. A year before *The Provok'd Husband* was staged, a pseudonymous letter appeared in *London Journal* under the name "Gracian." The letter was written in support of the signing of the Treaty of Hanover which had averted the threat of war with an Austro-Spanish alliance. The writer refers to those who supported a war against the alliance as having had a "miserable run of fortune" for there had been a number of events in the past decade or so that had promoted peace rather than war. For the warmongers Gracian advises,

I beg and conjure them, that instead of amusing themselves by computing the Odds and Chances of the Game [war] according to the Mathematick Laws of Du Moivre, they would in good earnest embrace the Advice of a *Spanish* Writer, and apply themselves to study the Humour and Complexion of their Fortune; for as *Fortune*, like other Ladies, can refuse him nothing to whom she has often been kind, so she continues obstinate never to grant any considerable Favours to those whom she at first beholds with Aversion and Contempt. [4]

Gracian knew vaguely of De Moivre's work, although he could not quite spell his name correctly. In his own mind, Gracian had not banished the goddess Fortuna as both De Moivre and Montmort wanted to happen as a result of their work. In 1731, another anonymous writer put De Moivre in a gambling context. The writer refers to gamblers who use mathematics in their approach to playing games of chance as "de Moivre men."[5] De Moivre's probably unintended impact on gambling continued throughout his lifetime. Another pseudonymous writer, Nestor, published an article in 1749 entitled "On contentment and avarice." When referring to a gamester, Nestor writes, "*Fortune* is his goddess, *De Moivre* his guide."[6]

How knowledge of De Moivre's work spread well beyond his own fairly large circle is difficult to say precisely. De Moivre financed the printing of his book by subscription. Presumably, this was done by contacting people whom he knew. He then contracted with William Pearson to print *Doctrine of Chances*. Once printed, he distributed the copies himself at Slaughter's Coffeehouse.[7] As to promoting the book, there were no advertisements from any booksellers in London newspapers offering the book at their stalls. The few mentions of *Doctrine of Chances* in the newspapers during the 1720s and early 1730s are when it was an item in an estate sale.

In one sense, Pearson as printer was an odd choice. Pearson's shop was close to a two-mile walk from where De Moivre lived. Furthermore, Pearson's main activity was in printing music,[8] although he did print a small number of mathematics books. It may have been William Jones who steered De Moivre towards Pearson. Jones chose Pearson to print Newton's *Analysis per Quantitatum Series, Fluxiones, ac Differentias* in 1711.

The printing and sale of *Doctrine of Chances* was very different from De Moivre's earlier tract against George Cheyne, the 1704 *Animadversiones in G. Cheynaei Tractatum de Fluxionum Methodo Inversa*. This book was printed by Edward Midwinter and then sold by Daniel Midwinter and his partner, Thomas Leigh, from their shop by St. Paul's Cathedral. Midwinter and Leigh were among the top booksellers in London.[9] Probably as a result of seeing someone else publish and sell Cheyne's book, they saw a potentially strong demand for books about the emerging new mathematics. In addition to De Moivre's book, in 1704 they sold copies of Charles Hayes's *A Treatise of Fluxions*. This was the first comprehensive treatment in English of calculus in a single publication to hit the market following Cheyne's Latin book the year before.[10] Edward Midwinter was also the printer.

Midwinter and Leigh also promoted the books they sold. They put out newspaper advertisements for *Animadversiones* and *A Treatise of Fluxions* when the books were available at their bookshop.[11] It was a good business decision. They carried one book by a fellow of the Royal Society that trashed the contents of the first calculus book to be published and at the same time offered a reliable alternative to the discredited book. Daniel Midwinter was still active as a bookseller, with several titles in mathematics in his shop, when *Doctrine of Chances* was first published. De Moivre could have left his new book in very capable hands, but instead decided to take complete control of the sale and distribution of it.

No known subscription list to *Doctrine of Chances* currently exists. Despite this, we can piece together a short list of some of the likely subscribers. De Moivre's good friends Edmond Halley and Isaac Newton, as well as friend and former pupil Martin Folkes, all had copies of the book in their libraries when they died.[12] Brook Taylor must also have had a copy since he was able to reply in detail to Pierre Rémond de Montmort's scathing comments about the book. Montmort was not a subscriber since he kept asking Taylor when the book would be available for purchase.[13] His copy may have come from Taylor or from some other subscriber. Like the subscription list to *Miscellanea Analytica*, some subscribers may have paid for more than one copy—not for themselves but to provide patronage to an important scientific publication. There are a couple of likely subscribers, one on the Whig political side and the other on the scientific but non-mathematical side. A copy of *Doctrine of Chances* was in Horace Walpole's library.[14] Since he claimed to have no interest or ability in mathematics and he was only a one-year-old when the book was published, the copy may have come from his father, Robert Walpole, who also subscribed to *Miscellanea Analytica*. Another copy shows up on a list for the sale of the library of John Chamberlayne in 1724.[15] Chamberlayne was a literary editor and fellow of the Royal Society. A relatively late arrival to obtain a copy of *Doctrine of Chances* is the Scottish mathematician James Stirling. From about 1717 until perhaps 1722, Stirling was in Venice, so it is unlikely that he was an original subscriber.[16] De Moivre did give a copy of the book to Stirling with a handwritten list of errata that he had drawn up.[17] Together, these provide a hint about how De Moivre circulated his book. A likely scenario is that the subscriptions covered the printing costs of the subscribers' books plus some or several extras. De Moivre kept the extra copies, including those not claimed by the subscribers, and sold them or gave them away over several years until his supply ran out.

By the early 1720s, a copy of *Doctrine of Chances* had found its way to Trinity College Dublin. There it was read by a twenty-eight-year-old undergraduate named Richard Dobbs.[18] Dobbs had a solution to the division of stakes problem when there are more than two players in the game. De Moivre had solved the problem in 1711 under Problem 8 in *De Mensura Sortis* and repeated his solution in *Doctrine of Chances*, also under Problem 8. This was a solution for which Montmort claimed priority and also recognized as cumbersome. Dobbs was aware of De Moivre's solution in *Doctrine of Chances* but, as he explained in a letter to the Royal Society,[19] his own solution "is rendered somewhat more easy & fit for practice." In his capacity as secretary to the Royal Society, James Jurin asked De Moivre to review Dobbs's manuscript. Dobbs's new solution is general and detailed, with tables to aid in the calculation of the required probabilities. It is also quite complex and at times difficult to follow. After receiving the manuscript, De Moivre sent it back to Jurin saying that Dobbs's method appeared correct, but that he had obtained an easier solution.[20] The reply letter from Jurin to Dobbs is only a draft and does not contain De Moivre's new method of solution.[21] In Dobbs's reply to Jurin, he says that De Moivre's solution "seems to be a considerable improvement of that he formerly published." Although it cannot be verified, De Moivre's new solution might be the one that shows up as Problem 69 in the second edition of *Doctrine of Chances* published in 1738.[22] It is certainly easier to implement than his original solution.

Throughout the 1720s, De Moivre continued to work on problems related to probability, as well as some other topics in mathematics. Some of the work he undertook was suggested by others in the same way that Francis Robartes had given De Moivre his lawn bowling problems and Thomas Woodcock his generalization to the gambler's ruin problem. Other work resulted from building on mathematical results that he had previously obtained and then applying these results to current problems that other mathematicians were working on. De Moivre collected much of this work together and published it in *Miscellanea Analytica* in 1730. With one exception, here I will concentrate on some of the work that can be identified to have originated in the mid-1720s or before. The one exception is the normal approximation to binomial probabilities. This work originates in a question posed to De Moivre in 1721. There are major developments into the 1730s and so it seems better for this topic to keep all the material together.

De Moivre's discovery of the use of generating functions dates from the mid-1720s. In this case it is difficult to say, although I will conjecture, what motivated him to revisit a problem that he had already solved in a reasonable way as early as 1711 using mathematical induction. The problem is to find the number of chances to obtain a particular sum of the numbers that show on the faces of several dice when they are thrown.

The problem appears in both *De Mensura Sortis* and *Doctrine of Chances* as a lemma inserted between problems on the number of trials required to obtain, with probability 1/2, at least one success, at least two successes, and so on, in a series of independent trials whose possible outcomes are either success or failure. De Moivre provides no proof of his result in his early publications, but outlines his method using combinatorial arguments and induction in a 1718 letter to Brook Taylor.[23] The proof by the use of generating functions first appears in the 1730 *Miscellanea Analytica*. Based on manuscripts held in the Macclesfield Collection at Cambridge University Library, there is evidence that the result by generating functions was likely obtained between 1722 and 1727.

Keeping the notation used for the lemma as it was described in Chapter 5, what is required is to find the number of chances to obtain the sum s on the faces that show in the throw of n dice each with f faces. Here is one way to look at the problem: Denote the die by D and make the exponent of D the value that shows on the face of the die; D^3, for example, is a die showing the number 3 on its face. The variable D is now an object rather than something that takes on numerical values. The sum $D^1 + D^2 + D^3 + ... + D^f$ expresses all the outcomes of a single die and the coefficient of each term in the sum is the number of ways that the face on the die can show—one way, in this case. For n dice, $(D^1 + D^2 + D^3 + ... + D^f)^n$ expresses all the outcomes of the dice. Expand this expression and collect the terms which have Ds with the same exponent. Since exponents are added when terms are multiplied together in the expansion, the term D^s is associated with s as the sum of the faces that show on the dice. The coefficient of D^s in the expansion is the number of outcomes for which the sum s shows on the dice. Since what is contained in the brackets of the expression is a geometric progression, it can be re-expressed as

$$\frac{D^n(1-D^f)^n}{(1-D)^n}.$$

This provides a generating function that, when expanded, can be used to obtain the number of chances associated with any sum s on the faces that show on the dice. It is easier to expand than the original form of the geometric progression. De Moivre's generating function does not include the term D^n and so his published generating function is

$$\frac{(1-D^f)^n}{(1-D)^n}.$$

When this simpler expression is expanded, the coefficient of D^{s-n} provides the number of chances associated with the sum s.

De Moivre essentially used the dice model I have given to explain his generating function. Instead of D^3, for example, to denote a die with the number 3 on its face, De Moivre used D^{111}. In order to get to the simpler generating function, he muddied the explanation and possibly confused his readers. He started with $1 + D^1 + D^2 + ... + D^{f-1}$, expressing all the outcomes of the throw of one die, which does not conform exactly to his model die that precedes this sum in *Miscellanea Analytica*.

This confusion also appears in two copies of a manuscript whose content can be attributed to De Moivre.[24] The copies are in the Macclesfield Collection held at Cambridge University Library. In the manuscript, De Moivre starts with a number of results related to basic counting rules. In terms of the notation that is given here, one of De Moivre's results on counting is that the sum of all dispositions of D quantities, "taking them by one's, two's, three's &c. to n" (or now f in the current notation), is given by the sum $D^1 + D^2 + D^3 + \ldots + D^f$. At this point he has used D as a quantity rather than as an object. Immediately following this in the manuscript, without any indication as to why, De Moivre changes to the sum $1 + D^1 + D^2 + \ldots + D^{f-1}$ and proceeds to give his version of the generating function. The manuscript finishes with an algebraic derivation of the formula for the number of chances to throw the sum s with the dice, followed by a numerical example to find the number of chances of throwing $s = 17$ points with four throws of a six-faced die.

The 1718 letter from De Moivre to Brook Taylor explaining how he solved the lemma in *Doctrine of Chances* puts a definite lower bound on the date when De Moivre discovered his generating function approach. The publication of *Miscellanea Analytica* in 1730 puts a definite upper bound on the date. The interval can be reduced somewhat by examining the manuscript result containing work on the generating function that is held in Cambridge University Library and then by looking at the historical context.

The two copies of the manuscript in Cambridge University Library have the title "Combinations," and are copies of the same material, one a rough copy in the hand of De Moivre's friend William Jones and the other done in a very fine hand.[25] The manuscript copy that is written in a fine hand looks as if it could be a presentation copy. It can be dated approximately. Not only is a fine hand used, but the paper is also of high quality. The watermark on the paper is that of Lubertus Van Gerrevink, a Dutch papermaker. It is the same as the watermarks numbered 317 and 318, both from Gerrevink, in a catalog of watermarks.[26] Van Gerrevink obtained patents for his watermarks in late 1726 and early 1727.[27] After he obtained his patent, Gerrevink used his distinct watermark to differentiate his paper from his competitors'. Several papermakers used Van Gerrevink's initials, LVG, as watermarks to associate their products with high quality paper. Van Gerrevink's patented watermark was unique to his paper. In the Macclesfield Collection, another manuscript set contains pieces of paper with the Van Gerrevink watermark.[28] The manuscript set bears the title "1733/1734/Annuities Upon Lives" written by William Jones, the owner of the manuscripts. It may be reasonably concluded that the manuscript written in a fine hand may be dated to between 1727 and 1734 approximately.

The copy in the fine hand was probably taken from the rough copy. It should not be concluded that the rough copy, or that the original (possibly a notebook kept by De Moivre) from which the rough copy was taken, dates from the late 1720s or later—my dating for the manuscript in the fine hand. Also in the Macclesfield Collection are two copies of a Newton manuscript related to probability entitled "Reasonings Concerning Chance."[29] There is both a rough copy in Jones's hand on the same type of paper as the rough copy entitled "Combinations" and a copy in a fine hand on Van Gerrevink paper. Again, no authorship or date is given on either manuscript. The original material from which these copies are taken is in Newton's notebook that he kept to record his expenses and to make notes on books that he had read while a student at Cambridge.[30] The relevant part of the notes also has the title "Reasonings Concerning Chance" and was likely written in 1665.[31] It contains notes based on Newton's reading of Christiaan Huygens's *De ratiociniis in ludo aleae.* Based on other information about Newton and De Moivre, a tentative dating of the rough copies of "Combinations" and "Reasonings Concerning Chance" could be narrowed to between 1722 and 1727.

The upper bound of 1727 is the year of Newton's death. There was dispute over his manuscripts.[32] Many of his surviving relatives wanted to benefit financially from his papers by publishing them. A review of the manuscripts by the Royal Society fellow Dr. Thomas Pellett determined that only two or three of them were publishable. Subsequently, John Conduitt, Newton's successor as Master of the Mint and husband to Newton's niece Catherine, posted a £2000 bond and obtained control of all the remaining manuscripts. Conduitt's unfulfilled ambition was to write a biography of Newton and so very few had access to the manuscripts during Conduitt's lifetime. After his death and that of Catherine, the manuscripts passed through the family and remained in private hands until well into the nineteenth century.

The lower bound of 1722 comes from De Moivre's correspondence with James Jurin over the manuscript sent to Jurin by Richard Dobbs concerning the division of stakes problem for more than two players. After the publication of *Doctrine of Chances* in 1718, De Moivre continued to work on probability problems. His initial work on approximating binomial probabilities for a large number of trials began in 1721.[33] The generating function approach to find the probability of the sum of the faces that show on dice involves multinomial expansions, rather than binomial ones. Dobbs refers to getting the solution of the division of stakes problem by expanding the multinomial of the form $(a + b + c + ...)^n$. After De Moivre read Dobbs's manuscript, he wrote to Jurin saying he had an easier solution. Having to think about multinomial expansions may have led De Moivre to consider his lemma in terms of what appears in the manuscript "Combinations." The multinomial was not a new topic for De Moivre. In his second published paper that appeared in 1697, De Moivre extended the binomial expansion to a multinomial one.[34] Specifically, in the current notation, De Moivre provides an expression for the expansion of a multinomial of the form $(aD^1 + bD^2 + cD^3 + ...)^n$, an expression that yields both Dobbs's multinomial and De Moivre's generating function as special cases. To make the final connection, De Moivre only needed to think of some of the terms in the multinomial as objects. Dobbs's paper may have stimulated De Moivre to think about the multinomial as a generating function.

The historical evidence provides no indication of how Jones came by the material that he copied into the manuscripts "Combinations" and "Reasonings Concerning Chance" and why he, or someone else, made fair copies of each. That is a subject of reasonable conjecture. Since De Moivre and Jones were close friends,

I would suggest that De Moivre showed Jones some papers or a notebook in his possession from which Jones made his rough copy of "Combinations." This would be in line, for example, with De Moivre informing Brook Taylor about some of his latest work. De Moivre and Newton were also close friends, meeting in Slaughter's Coffeehouse in St. Martin's Lane, a very short walk from the house that Newton occupied after 1710. Matthew Maty reports that after finishing at Slaughter's, Newton and De Moivre often went to Newton's house, "where they spent their evenings debating philosophical matters."[35] A likely scenario is that around the time De Moivre dedicated the first edition of *Doctrine of Chances* to Newton, Newton showed De Moivre his college notebook containing his notes on Huygens. Through De Moivre, Jones gained access to the notebook and again copied out the relevant parts. In the late 1720s or early 1730s Jones commissioned fair copies of the two manuscripts "Combinations" and "Reasonings Concerning Chance." These were done on the high quality Van Gerrevink paper. They were commissioned as some kind of presentation copy after the death of Newton in 1727 and after the methodology of generating functions had been revealed in *Miscellanea Analytica* in 1730. Since Jones had tutored George Parker, 2nd Earl of Macclesfield, during the earl's youth, as did De Moivre, and since Jones regularly stayed at Shirburn, the Macclesfield country seat, the presentation copies were probably made for the earl, who was a subscriber to *Miscellanea Analytica*. Of course, other possibilities could be constructed. The one I have given seems to me to be the simplest.

De Moivre kept mum on the new approach via generating functions until he published it in *Miscellanea Analytica*. A few of his close friends may have known about the new way to solve the lemma for the sum on the dice. William Jones and James Stirling are two possibilities. Jones's copy of "Combinations" was almost certainly in his hands before Newton's death in 1727. The argument for Stirling requires a little more detail. Near the end of the 1720s there was some renewed interest in De Moivre's lemma that so far had only appeared without proof. Gabriel Cramer, a Swiss mathematician at the University of Geneva, began wondering about proofs for the lemma in 1727.[36] The next year, Cramer was in London visiting various mathematicians and other scientists. At about this point he heard about De Moivre's forthcoming publication of *Miscellanea Analytica* and arranged to have a subscription to it. After he left London, Cramer found his own solution to the lemma using mathematical induction and sent it to James Stirling in a letter dated October 11, 1728.[37] Stirling replied that he had seen De Moivre's solution and Cramer wrote back on March 12, 1729, asking if his own solution agreed with De Moivre's.[38] The letters from Stirling to Cramer are not extant.

With respect to De Moivre's generating function, there are some precedents to thinking of terms in an algebraic expression as objects that leads to this kind of function. And they go back to some of the earliest work in probability theory. The use of the binomial expansion as a generating function is implicit in Pascal's 1665 work on the arithmetical triangle.[39] A simple explicit example, closer to the time when De Moivre was writing, is in John Arbuthnot's work on divine providence in 1710. When trying to calculate the chances of obtaining an equal number of male and female births Arbuthnot begins with the following discussion:

> Let there be a Die of Two sides, M and F, (which denote Cross and Pile), now to find all the Chances of any determinate Number of such Dice, let the Binome M + F be raised to the Power, whose Exponent is the Number of Dice given; the Coefficients of the Terms will shew all the Chances sought. For Example, in Two Dice of Two sides M + F the Chances are $M^2 + 2MF + F^2$, that is, One Chance for M double, One for F double and Two for M single and F single.[40]

In his argument, the terms M and F are objects. Assuming that male and female births occur with equal probability, the coefficients in the binomial expansion give the number of chances for the various numbers of male and female births that could occur. With a total of four births, the coefficient of the object M^1F^3, for example, in the expansion is the number of chances of seeing one male and three females born.

One result that De Moivre obtained in the early 1720s is not about probability, but rather his work surrounding the semicircle used to find a trigonometric solution to the duration of play problem. It also brings De Moivre back to an old favorite topic: quadratures. De Moivre's work was motivated by a problem that had been partially solved by Roger Cotes, a mathematician who held the Plumian Chair of Astronomy at Cambridge. Cotes was interested in finding the fluents of certain fluxional quantities. He died in 1716, leaving a number of unpublished manuscripts. They were eventually collected and edited by his cousin and successor in the Plumian Chair, Robert Smith, who published the work in 1722 under the title *Harmonium Mensurarum*.[41]

Through a challenge put out by Brook Taylor late in 1718, initially to Montmort,[42] the fluxional quantity that became of interest, both in England and on the Continent, is one of the form

$$\frac{\dot{x}x^{(d/n)q-1}}{a+bx^q+cx^{2q}},$$

where a, b, c, and q are constants, d is a nonzero integer, and n is a power of 2. Finding the fluent of this quantity is equivalent to finding the quadrature, or the integral, of the curve

$$y = \frac{x^{n-1}}{a+bx^n+cx^{2n}};$$

this is accomplished through finding the quadratic factors of the denominator. De Moivre was able to find these factors for the case when $a = c = 1$ and $b = -2\cos(\theta)$, where θ is any angle so that b is any number between -2 and 2. He also generalized the result to any positive integer n rather than a power of 2. Although the result appears in the 1730 *Miscellanea Analytica*, James Stirling informed Nicolaus Bernoulli in a letter written in 1730 that De Moivre had obtained his factorization shortly after Cotes's book was published in 1722.[43] De Moivre's interest was probably piqued by Cotes's description, via Smith as editor, that part of the solution to a similar problem has a circle divided into $2n$ equal parts.[44]

In order to obtain his factorization, De Moivre began with a result in *Miscellanea Analytica*, which in modern notation is given by

$$\cos(\theta) = \frac{1}{2}\left(\cos(n\theta) + i\cdot\sin(n\theta)\right)^{1/n} + \frac{1}{2}\left(\cos(n\theta) - i\cdot\sin(n\theta)\right)^{1/n},$$

where i is the imaginary number $\sqrt{-1}$. It is stated without proof. An equivalent result, though not given in any of De Moivre's work, is

$$\left(\cos(\theta) + i\cdot\sin(\theta)\right)^n = \cos(n\theta) + i\cdot\sin(n\theta).$$

The latter result is now known in mathematics as De Moivre's theorem or De Moivre's formula. Although it appears to come out of thin air, De Moivre probably obtained his initial equation in *Miscellanea Analytica* as early as 1707. As mentioned in Chapter 5, De Moivre had a method to find an algebraic expression for one of the roots of a certain polynomial equation in 1707 and provided details of his method in 1722.[45] In the 1722 paper, he gives two equations related to his *Miscellanea Analytica* result: $1 - 2z^n \cdot \cos(n\theta) + z^{2n} = 0$ and $1 - 2z \cdot \cos(\theta) + z^2 = 0$. The solutions to the first equation are $z^n = \cos(n\theta) \pm i \cdot \sin(n\theta)$ or $z = (\cos(n\theta) + i \cdot \sin(n\theta))^{1/n}$ and $z = (\cos(n\theta) - i \cdot \sin(n\theta))^{1/n}$. From the second equation there are two other solutions for z: $\cos(\theta) \pm i \cdot \sin(\theta)$. Adding together these last two solutions for z yields $z = \cos(\theta)$. De Moivre's formula follows from adding the two solutions for z from the first equation and equating the result to the solution for $2z$ from the second equation.

De Moivre continued to work on, or at least promote, areas of mathematics that were only very loosely related to his work in probability. He was interested in the practical aspects of logarithms, especially natural or hyperbolic logarithms. Tables of hyperbolic logarithms would have been useful to him for his two approximations to the binomial—normal and Poisson. In the late 1720s there was a movement afoot to construct tables of natural logarithms, though it apparently never came to fruition. The project got at least as far as De Moivre writing the preface to the proposed set of tables. The manuscript version of the preface is in the Macclesfield Collection.[46]

In the preface, De Moivre gives the motivation for constructing tables of natural logarithms as filling a practical need in natural philosophy. Many problems in natural philosophy depend on the quadrature of curves, the most useful being the circle and the hyperbola. Within these two types of curves, the most convenient to use are the unit circle defined by $x^2 + y^2 = 1$ and the equilateral hyperbola that can be put in the form $y = 1/x$. The first quadrature can be found from tables of sines and tangents, while the second requires tables of logarithms. Although extensive tables of common (base 10) logarithms were available, tables of hyperbolic or natural logarithms would be much more convenient. To illustrate his point, De Moivre took a numerical example from page 344 of the 1726 edition of Newton's *Principia* and showed how much easier the answer was to obtain with natural logarithms. Since De Moivre refers to this edition of the *Principia* as the "new" edition and makes no mention of Andrew Motte's 1729 translation of the *Principia*, the new table of hyperbolic logarithms was probably intended for publication between 1726 and 1729. And the author of the intended table was probably De Moivre's friend William Jones since the manuscript version of the preface was in his possession. In his introduction to *The Anti-Logarithmic Canon* published in 1742,[47] James Dodson outlines the history of the development of logarithms and the calculation of tables of common logarithms with no mention of extensive tables of natural logarithms. This was a development that would have to wait for several years to come to pass.

❧ 10 ❧

De Moivre as Teacher

The social background from which De Moivre's students came dictated the topics that they covered in their lessons. His students were mainly young "gentlemen," which means, in the eighteenth-century use of the word, that they were from landed families or from families of some rank or distinction. We can get a general picture of the type and level of mathematics that De Moivre taught from a 1745 newspaper advertisement for the mathematics curriculum at an academy at Heath near Wakefield in Yorkshire.[1] The academy was privately run and had no formal ties to any Christian denomination, unlike the Huguenot academies in France that De Moivre attended or some other schools and academies in England. The curriculum for a gentleman at the Heath Academy was, "A course in mathematics and philosophy, viz. geometry, geography, astronomy, and natural philosophy; the valuation of estates, annuities and reversions." A gentleman, to live as a gentleman, must therefore be versed in the ideas of the new sciences as well as the practical aspects of his day-to-day life related to a landed estate. The study of natural philosophy included topics in mechanics, hydrostatics, pneumatics, and optics. As a foundation for this study, the student would have to know arithmetic, algebra, Euclidean geometry, and conic sections.

As noted in Chapter 3, according to one eighteenth-century classification of mathematics, topics in the field divided themselves into two general areas: pure and mixed mathematics.[2] Arithmetic, algebra, geometry, trigonometry, and conic sections belong to pure mathematics; topics such as geography, astronomy, optics, and mechanics belong to mixed mathematics. De Moivre taught his students topics in pure mathematics.

Some younger sons of the landed classes took the route of entering the army or navy as a profession. They needed mathematics for gunnery and navigation.[3]

The pure mathematics part of the curriculum at the Heath Academy for "gentlemen of the Army" and "gentlemen of the Navy," i.e., for those destined to be officers, included geometry and trigonometry. At least one De Moivre student, Edward Montagu's younger brother John, followed a career into the army, rising to the rank of lieutenant-colonel in the Foot Guards.[4]

The major centers of study for gentlemen in the eighteenth century were the universities at Oxford and Cambridge. There was a religious stumbling block to study at these universities; students had to be willing to subscribe to the Thirty-Nine Articles of Faith as laid down by the Church of England. Those who did not conform to the Thirty-Nine Articles—so-called nonconformists or dissenters—were barred from being admitted to Oxford and were not allowed to graduate from Cambridge. Consequently, some dissenters set up their own academies in parallel to Oxbridge that provided some gentlemen scholars with another route to education. Study at Oxford and Cambridge could be used as preparation for the Anglican ministry; the primary focus of the dissenting academies was for preparation of students for ministry in the dissenting or nonconforming churches, such as those attended by Presbyterians, Congregationalists and Baptists. Lay students also attended these dissenting academies.

The mathematics curriculum offered at Oxbridge and at the dissenting academies was similar. In the early eighteenth century, two different tutors at Cambridge, one from Clare College and the other from Magdalene College, described the course of study students should follow at university. Robert Greene's students at Clare College began the study of mathematics in their second year at university.[5] His curriculum from 1707 shows students studying arithmetic, geometry, and algebra in their second year, conic sections in the third year, and fluxions, logarithms, trigonometry, and infinite series in the fourth year. The treatment of fluxions and infinite series at this time was unusual, so it is difficult to say if any of Greene's students ever advanced to this material. The pure mathematics topics were studied alongside topics in astronomy, optics, and mechanics. Daniel Waterland began his career as a tutor at Magdalene College and rose to the position of master of the college. His syllabus circa 1730 shows students beginning their mathematical work in their first year of study with arithmetic, geometry, and trigonometry.[6] Conic sections followed in the second year. After second year, some suggested topics of study included astronomy, physics, and optics. The dissenting minister Philip Doddridge has described his course of studies between 1719 and 1723 under John Jennings at the Kibworth Academy.[7] The mathematical subjects taught at this dissenting academy were arithmetic, algebra, and geometry in the first year and then selected topics from mechanics, hydrostatics, physics, and astronomy in the second year. Doddridge continued and built on this curriculum in the academy that he opened at Northampton in 1729. At another dissenting academy, the first-year mathematics topics included arithmetic, geometry, and trigonometry. The topic of conic sections was covered in the second year. Other topics through the second to fourth years include optics, physics, and astronomy.[8] The surviving evidence from De Moivre, as well as from his friend and fellow tutor William Jones, is that they taught material at a level comparable to what was taught at the universities and dissenting academies. In view of the family backgrounds of his students, De Moivre would have kept his eye almost exclusively on the universities rather than on the dissenting academies when preparing his students.

Here is a list of De Moivre's students separated into "definite" (ones that can be confirmed as his students based on source material) and "probable" (based on arguments made though an analysis of the subscription list to *Miscellanea Analytica*).[9]

- Definite
  - ✧ Charles Cavendish
  - ✧ Peter Davall
  - ✧ James Dodson
  - ✧ Martin Folkes
  - ✧ Edward Montagu
  - ✧ John Montagu (2nd Duke of Montagu)
  - ✧ George Parker (2nd Earl of Macclesfield)
  - ✧ Michel de la Roche
  - ✧ Philip Stanhope (2nd Earl of Stanhope)
  - ✧ George Lewis Scott
  - ✧ Henry Stewart Stevens

- Probable
  - ✧ James Cavendish
  - ✧ William Cavendish (2nd Duke of Devonshire)
  - ✧ Richard Edgecomb
  - ✧ Francis Fauquière
  - ✧ Coulson Fellowes
  - ✧ Martin Fellowes
  - ✧ William Fellowes
  - ✧ William Folkes
  - ✧ George Furnese
  - ✧ Henry Furnese
  - ✧ Isaac Guion
  - ✧ Colonel John Montagu
  - ✧ William Montague (2nd Duke of Manchester)
  - ✧ Henry Pelham
  - ✧ Thomas Pelham-Holles (1st Duke of Newcastle-upon-Tyne)
  - ✧ Thomas Townshend
  - ✧ William Townshend
  - ✧ Edward Walpole

Of the eleven individuals who definitely studied with De Moivre, only three attended university, all at Cambridge. Of the eighteen probable students, half of them went to university, mostly to Cambridge with a few to Oxford.[10] The remainder received a gentlemanly education strictly through private tutoring or some other route. As to the age of De Moivre's students, the historical evidence points to De Moivre taking on students when they were about sixteen years of age or slightly older.[11] There are exceptions to this rule such as the case of Michel de la Roche. Also Philip Stanhope was prevented from studying mathematics by his uncle and guardian, Philip Dormer Stanhope, 4th Earl of Chesterfield; the young Stanhope's studies in mathematics began only when he became independent from his uncle.[12]

De Moivre had lodgings, not a house. He needed a place to carry out his work as a tutor. He also taught gentlemen whose fathers probably expected certain privileges based on their rank. The situation is illustrated by another mathematics teacher, John Ward who may have been one of De Moivre's competitors for students. In 1695, Ward advertised that he taught at home and at the premises of a mathematical instrument maker.[13] In addition, Ward's advertisement says that "the nobility and gentry are taught at their own houses." The latter situation describes De Moivre's teaching career exactly. In a 1707 letter to Johann Bernoulli,[14] De Moivre described his teaching day. He taught from morning until night and had to walk to where his students lived in order to give them instruction. He must have taught several students in a single day in various parts of the city since he told Bernoulli that much of his time was spent walking around London.

Currently, only one set of notes from one of De Moivre's students is known to survive.[15] Written in French, they are on arithmetic and algebra, and come from a time that is late in De Moivre's career. The notes were the property of Friedrich Georg Brandes. He tutored Georg Friedrich von Steinberg, who came from a distinguished family in Hanover. Brandes took his pupil on the Grand Tour; they were in London during 1742 and 1743. The lessons on which the notes were based occurred between May 7, 1742, and the end of April 1743. The manuscript is written in two different hands. In one hand there are notes on algebra, and in the other hand there are notes on arithmetic. I have compared the handwriting in the arithmetic part of the notes to De Moivre's letters to Philip Stanhope in the 1740s.[16] The handwriting on both appears very similar.

In the arithmetic section, the operations of addition, subtraction, multiplication, and division are covered for fractions and decimals. This is followed by material on the extraction of squares roots, calculations for compound interest and annuities, and arithmetic using British currency with pounds, shillings, and pence. The part on algebra has material on the solution to equations in two and three unknowns, arithmetic and geometric progressions, and the solution to quadratic equations. There is a large overlap of this material with Philip Doddridge's lecture notes on algebra for the course he taught at the Northampton Academy.[17] De Moivre's notes contain some additional material, including a few problems in recreational mathematics, some elementary number theory problems, such as finding the integral solutions to the equation $x + y + x \cdot y = 11$ and to $x^2 + y^2 = 13$, and problems related to the application of algebra to geometry, such as calculations related to right-angled triangles.

Here is one example of De Moivre's problems in algebra that he set for his student: Two people, A and B, are 59 miles apart. Person A travels 7 miles in 2 hours, while B goes 8 miles in 3 hours. B sets out one hour after A. How far, in miles, does A go before he meets B? To answer the question, the student must equate the lengths of time until A and B meet. Let x be the required distance in miles travelled by A. Then the distance travelled by B is $59 - x$ miles. The length of time travelled by A is then $2x/7$ and the length of time travelled by B is $3(59 - x)/8$. Then equating the times until they meet yields $2x/7 = 1 + 3(59 - x)/8$, since B starts one hour later than A. The solution is $x = 35$ miles.

What is interesting about this problem, other than the level of mathematics taught, is that William Jones, who also worked as a tutor, set the same question for his students.[18] The problem originates in print in Isaac Newton's *Arithmetica Universalis*[19] which is based on his lectures from Cambridge between 1673 and 1683.[20] The question is from a section on how worded questions can be expressed as equations.[21] William Whiston, Newton's successor in the Lucasian Chair at Cambridge persuaded Newton to have his lectures printed. Both Whiston and his successor, Nicholas Saunderson, taught from *Arithmetica Universalis* while at Cambridge. From this, and the overlap with Philip Doddridge's lecture notes, it may be inferred that De Moivre was tutoring mathematics at least at a level corresponding to a student in his early years at Cambridge or at the dissenting academies.

In his youth, De Moivre learned his arithmetic from François le Gendre's *L'Arithmétique en sa perfection*. Common to books on commercial arithmetic of this time, many arithmetical rules are followed by practical examples, often taken from commercial settings. In his notes on arithmetic, De Moivre's approach is similar to le Gendre's. For example, both le Gendre and De Moivre discuss the rule of three. Three numbers a, b, and c are given and it is required to find a fourth number x through the relationship $a/b = c/x$. Le Gendre begins his discussion with this question: If 24 men have supplies that will last 12 days, how long will supplies last for 15 men? This question is followed by several variations on the same question using different numbers.[22] De Moivre's numerical example in this case is, if 18 yards (verges) of something cost £32, what do 25 yards cost? When it comes to the topics of multiplication and division of fractions, both le Gendre and De Moivre show how to carry out the operations from a given set of fractions. Numbers only are given without any commercial context.

With respect to teaching material on interest and annuities, the amount of material in the arithmetic notes is small. And it is very elementary. De Moivre gives a numerical example of finding the interest payable for a fraction of a year assuming simple interest. This is followed by another numerical example.

Using longhand multiplication, the accumulated values after two, three, and four years of an amount are found under the assumption of compound interest. De Moivre's last set of problems in interest and annuities is to find the present value of an annuity of £1 per year for "1, 2, 3 &c." years at 4% interest. He carries out the evaluation of $1/1.04 + 1/1.04^2 + 1/1.04^3$ by first giving the values for the quantities $1/1.04$, $1/1.04^2$, and $1/1.04^3$ without calculation. This is followed by a longhand addition of those values. The material on annuities is concluded with a problem to find the present value of an annuity of £325 for twenty years at 4% interest. What simplifies the question enormously is that De Moivre gives, without calculation, the present value of an annuity of £1 for the twenty years.

No other teaching material on annuities that can be attributed to De Moivre has survived. We can get an idea of what else he might have taught by looking at his friend and fellow tutor, William Jones. Jones's teaching materials are held in the Macclesfield Collection at Cambridge University Library. The collection contains notebooks on simple and compound interest, as well as notebooks on annuities.[23] The material covered is similar to what appears in commercial arithmetic books. The typical flow of the material is that some general theory is given, followed by worked examples. From some of Jones's correspondence, at least some of the worked examples are taken from real situations.[24] Jones's general approach to annuities is in line with De Moivre's notes on algebra that were owned by Brandes. Jones's notebooks have a more thorough coverage of the topics.

De Moivre subscribed to a 1731 textbook on conic sections written for students at Cambridge.[25] The vast majority of subscribers to the book were Oxbridge students or teachers. In view of his purchase of the book, De Moivre undoubtedly taught his more advanced students conic sections. His purchase of the book indicates that he wanted to know what his students needed to know in order either to prepare them for university or to cover the material that they would have taken had they attended university. In either case, his students would receive the necessary mathematical part of a gentlemanly education.

There are some indications that De Moivre taught material beyond the standard mathematics syllabus at Oxbridge and the dissenting academies. An 1874 catalog of French manuscripts in the St. Petersburg Library lists a manuscript that contains extracts of some lessons by De Moivre on probability.[26] It came from a large collection of manuscripts owned by Andrzej and Józef Załuski. The collection had been transferred from Poland to St. Petersburg in 1794 by Czarina Catherine II. After contacting the library, now called the National Library of Russia, I was informed that the De Moivre manuscript was no longer held by the library but was given its next destination. Since the original owners had founded the first public library in Poland in the eighteenth century, the Soviets returned the entire Załuski collection of manuscripts to the National Library of Poland in the late 1920s under the Treaty of Riga. Upon contacting the library in Poland, I was informed that the De

Moivre manuscript, along with most of the Załuski collection, was destroyed during the Second World War.[27]

Both sets of notes, the one on probability and the other on algebra, are, or were as the case may be, in French. It is safe to conclude that at least some of De Moivre's lessons were given in French. A sample of size two is insufficient to generalize this observation to describe his day-to-day teaching.

❧ 11 ❦

Life Annuities

We have already seen that when writing the preface to his uncle's *Ars Conjectandi* in 1713, Nicolaus Bernoulli encouraged Abraham De Moivre to work on some economic and political applications of probability. De Moivre was interested, but declined due to his teaching workload.[1] Five years later, in his preface to his *Doctrine of Chances*, De Moivre suggested that Nicolaus himself was better suited to work on these kinds of problems in view of his own work in the area. He also suggested that Johann Bernoulli was well qualified to do the work. With respect to Nicolaus, De Moivre was undoubtedly referring to a summary of his PhD dissertation, which appeared as a supplement to the *Acta Eruditorum* for 1711.[2] Part of Nicolaus Bernoulli's thesis is concerned with probabilities of survival in human populations, as well as the valuation of life annuities and life insurance.

With words such as this flowing from De Moivre's pen, it seems rather odd that seven years after the publication of *Doctrine of Chances* De Moivre published a short book on an economic application of probability.[3] It turned out to be highly influential. This was the 1725 *Annuities upon Lives* and no reference to any Bernoulli can be found in it. The two names that are mentioned in the book are Edmond Halley and Thomas Parker, 1st Earl of Macclesfield. Halley's name appears throughout the book; De Moivre used a model for the probability of survival that was inspired by a life table that Halley had constructed in 1693 and published in *Philosophical Transactions*.[4] It is to Macclesfield that the epistle dedicatory is addressed and only there does Macclesfield's name appear. The letter to Macclesfield opens, "I should not presume to inscribe the following piece to your Lordship, were it not that the subject it treats of has been made the entertainment of some of your leisure hours."

In knowing how the earl spent some of his leisure hours, De Moivre probably knew Macclesfield fairly well. Macclesfield had been a fellow of the Royal Society since 1712. De Moivre also tutored Macclesfield's son and heir, George.

The choice of Macclesfield as dedicatee is an interesting one for two reasons. The first is political. De Moivre's letter, addressed to "The Most Honourable Thomas Earl of Macclesfield, Lord High Chancellour of Great Britain," is dated January 1, 1725; the book appeared in the booksellers' stalls near the end of February.[5] Parker was one of the emerging "new men" of the late Stuart and early Georgian era. He rose from the ranks of the minor gentry to become Lord Chancellor in 1718. This was one of the top political appointments in Britain; the Lord Chancellor was the chief judge in the High Court of Chancery, which handled cases of equity. Parker was created Earl of Macclesfield in 1721. Then in an abrupt reversal of fortune, Macclesfield was removed from his office as Lord Chancellor on January 4, 1725. Problems for Macclesfield had been brewing since late 1724. There had been rumors that the masters in Chancery, clerks under the Lord Chancellor, had been misusing funds entrusted to them by suitors, and that Macclesfield had been encouraging this practice. The total amount of missing funds was estimated at £60,000, a staggering sum for the time. A commission had been appointed in November 1724 to investigate. After leaving office, Macclesfield was subsequently impeached and then fined £30,000 in May 1725.[6] Knowing within three days of writing it, or perhaps even prior to writing it, that Macclesfield was in serious trouble politically and perhaps financially, leaving the letter in was either an expression of De Moivre's political naïveté or his loyalty to someone who was close to him in some way. The second reason why De Moivre's choice of Macclesfield as dedicatee is interesting is best expressed in the form of a question: Why was Macclesfield interested in life annuities in the first place? When the answer to the question is fully unpacked, it reveals not only Macclesfield's motivation, but also the scenario for the future impact that the book would have on De Moivre's career and on the careers of several others.

To explore the Macclesfield question, it seems best to follow the King of Hearts' advice to the White Rabbit: begin at the beginning and go on till you come to the end. In this case, it means peeling away a number of historical layers dating back to 1692 and before, and then examining those layers up to about 1725. All the layers involve life annuities in one form or another. This excursion will take a little time. In exercising your patience with this excursion, what I hope to achieve is the reward of knowledge of the historical context in which De Moivre was writing his *Annuities upon Lives*. This will go some way in getting a handle on comments from historians such as Loraine Daston, who has written that in the first two thirds of the eighteenth century, mathematicians had little impact on the price of life annuities in the marketplace.[7]

There were two contracts involving annual payments that were contingent upon the survival of one or more individuals. The first and oldest was tied to the land, and is known generally as "leases for lives." A tenant could take out a lease on a piece of land where the term of the lease extended until the death of the lessee or the death of the last survivor of the lessee and up to two others named in the lease. The second contract is what we would normally consider today as a life annuity. A purchaser paid a lump sum of money—to the government or to a private company or to an individual—in order to receive an annual payment until the death of a named individual. The named individual did not have to be the purchaser of the annuity. Like leases for lives, there was also the possibility of joint-life annuities that made annual payments until the first death or until the death of the last survivor. Although these two financial ideas are quite different, they are mathematically equivalent. The value of the lease is the present value of all the rental payments, essentially annuity payments to the landlord. For a standard annuity, a fair purchase price is the present value of expected future payments to the annuitant. There is one difference in substance between leases for lives and joint-life annuities. In the leasing system, if one person dies, then, following the payment of a fine, that person could be replaced by someone else in the lease. The mathematics behind the evaluation of the fine has the same underlying principles as the evaluation of standard life annuities.

There were three general types of land tenure that carried into the eighteenth century: freehold, copyhold, and leasehold.[8] In a freehold estate, the purchaser held the property in perpetuity or until he sold it. Freehold estates were often entailed, meaning that that there was a specification for how the land passed from one owner to the next, often one generation to the next. For copyhold land, the lessee typically paid a fine for entry into the lease and then paid annual rents for the term of the lease. The payments could run for a fixed number of years or for the lengths of the lives of up to three people named in the lease. Traditionally, one life was valued at seven times the rental payment, referred to in the contemporary literature as "seven years purchase." Two lives were valued at fourteen years purchase and three lives at twenty-one years purchase.[9] In a leasehold estate, the lease might specify a fixed period or it might be based on the life of the lessee with the possibility that the lease would pass to the widow or other named individual on the death of the lessee. Evidence for the use of leases for lives dates from at least the time of Henry VIII; a 1541 statute stipulated that the maximum length of a lease was twenty-one years or three lives.[10] Based on this long-established system of payments, standard methods of compound interest could be used to calculate rents and fines as well as the values of estates or leases. By the early eighteenth century, the system of rents and fines for leases for lives was so entrenched that it was very slow to change even as the mathematics developed to allow a better valuing of the lease.

Given this system of land tenure, it was in the landlord's interest to be knowledgeable in the calculation of interest and fixed-term annuities. This would allow him to set rents and fines, to sell or purchase estates, and to interact fully with his steward who collected these moneys. Consequently, interest and annuities were common subjects studied by the sons of the landed classes. Practice questions on the valuing of estates and the setting of rents appear in many seventeenth-century commercial arithmetic books.[11] It would have been one of the topics that De Moivre taught to his students. Appearing on the subscription list to De Moivre's 1730 *Miscellanea Analytica* were a number of students from landed families, some of whose associations with De Moivre go back to the late seventeenth and early eighteenth centuries.[12] De Moivre's facility with fixed-term annuities is shown through his interaction with Edmond Halley on the subject in about 1706 when De Moivre verified a formula for Halley to determine the interest rate in any annuity.[13]

Like leases for lives, some forms of life annuities had been around for some time. Typically, it was land based. For example, on his death, a landlord might bequeath a life annuity to his widow. In order to do this, some land was set aside or alienated from the estate, and the rents from the alienated lands provide the annual payments. On the death of the widow, the lands reverted to the landlord's heir. This did not require any new or complicated mathematics. The sale of life annuities on the open market was a different matter. And the open market only developed in 1692 as a result of financial need on the part of the government.

The idea of selling life annuities to the public as a means of government finance was imported from elsewhere. When William of Orange along with his wife, Mary, were imported from the Low Countries in 1688 to become William III and Mary II, with them came some Dutch ideas of finance that had been practiced there since the sixteenth century.[14] The English government needed money, a lot of it, to finance the military campaigns of King William III and his allies against France. Campaigning began in 1689 and continued yearly during the summer fighting season. The conflict, known as the Nine Years' War, ended with the signing of the Treaty of Ryswick in 1697. The war consumed about 80% of British public revenues.[15] The use of the life annuity for British public finance was instituted in an act of 1692 that empowered the government to raise £1 million through the sale of these annuities.[16] Subsequently, the act was called the "Million Act" and the annuities were often called Exchequer annuities since they were sold through the Office of the Exchequer. In a sharp break from the past, Exchequer annuities were not secured by land. Rather, they were secured through an excise tax on beer and liquor. Further, it was an opportunity, completely new, for people to obtain a life annuity without having to own land.

It was at this point that Edmond Halley, by chance, walked onto the life annuity stage.[17] Caspar Neumann, the Lutheran pastor in the city of Breslau in Silesia, had collected data on the total number of births and the number of deaths at each age from the church registers of that city. The data covering the years 1687 to 1691 eventually found their way to the Royal Society for Halley to analyze. Regarding life annuities, Halley did three things with Neumann's data. First, he calculated a life table, or more correctly a population table, which gives estimates of the number of people alive in Breslau at each age up to 84. For some reason Halley provided no information for ages 85 and up other than the total number of people in that age group. Halley's table is probably the first life table ever to be calculated based on population data.

The second thing Halley did was to use the life table to calculate the present value of a life annuity at various issue ages. From reading the minutes of Royal Society meetings, one gets the impression that this was possibly an afterthought. Halley presented his life table to the Royal Society on March 8, 1693.[18] It was not until a week later that he presented his annuity calculations. The third thing Halley did was that he indicated how joint-life annuities, for two and three lives, could be calculated from his table.

In modern actuarial notation, a_x is the present value of a life annuity at rate of interest i of an amount 1, payable annually at the end of the year to a person aged x until death occurs. The value of this annuity is calculated from the formula

$$a_x = \sum_{t=1}^{\omega - x} v^t \frac{l_{x+t}}{l_x}.$$

In this expression, the term $v^t = (1 + i)^{-t}$ is the present value of an amount 1 to be paid t years in the future assuming a rate of interest i. The term l_x is the number of people alive at age x, so the quotient l_{x+t}/l_x is the probability that a person aged x survives to age $x + t$. The values of l_x are obtained from a life table, such as the one produced by Halley. The product $v^t l_{x+t}/l_x$ is the expected value of the payment at time t in the future, so a_x is the expected value of all future annuity payments. The survivor probability l_{x+t}/l_x is also denoted in the actuarial literature by the expression $_tp_x$. Finally, the Greek letter ω stands for the end of life, so $l_\omega = 0$. The life annuity might be compared to a standard annuity for a fixed term, say m years. The usual actuarial notation for the value of this annuity is $a_{\overline{m}|}$. It is calculated from the formula

$$a_{\overline{m}|} = \sum_{t=1}^{m} v^t.$$

In this case, the annuity payments are made with certainty, rather than being contingent on the survival of the annuitant.

Using his life table and tables of logarithms, and with paper and quill pen only, Halley calculated some values for a_x. He did this at a 6% rate of interest, the legal rate at the time, and at age $x = 1$ and the ages $x = 5$ quinquennially through 70. The burden of hand calculation for these valuations is enormous, so it is not surprising that Halley made a couple of numerical errors along the way.[19] In view of the burdensome nature of the calculations, it is also not surprising that Halley made no numerical valuations of annuities for joint lives.

Halley gave five uses for his life table. His fifth use, which appears almost as an afterthought, like his annuity calculations themselves, is to use his annuity valuations to comment on the Exchequer annuities offered under the Million Act of 1692. The annuities were priced at seven times the amount of the annual payment (the traditional seven years purchase for one life) regardless of the age of the purchaser. Halley commented that for many it was advantageous to buy these annuities. For those aged 65 or younger, a proper price should have been more than seven times the annual payment according to his annuity valuations based on the Breslau data. At least one person tried to capitalize on the bargain. In 1703, John Blunt, a London money-lender advertised that he would buy Exchequer annuities from those who possessed them, provided that the person named in the annuity was in good health.[20] Even a decade after they were initially marketed, Blunt offered 10% above the original purchase price.

After Halley's brief fling with life annuities, he exited this stage. As we shall see, he was not exactly "heard no more," but only heard of very little until 1725. His ideas remained before the public. Halley's paper was reprinted in 1705 and 1708 in collections of papers entitled *Miscellanea Curiosa* that were taken from *Philosophical Transactions*. Some of the papers were abridged from the originals in *Philosophical Transactions* and others were translations of the Latin originals.

After the government entered the life-annuity business, the floodgates soon opened. Some enterprising individuals decided to finance their business schemes by the sale of life annuities, where the only security was in the success of the enterprise.[21] For example, in 1714 the Oil Annuity Office commenced operation. It was set up to raise £20,000 through the sale of life annuities in order to finance a scheme to extract oil from the nuts of beech trees. Others decided to market annuities by alienating land to provide the security. Based on a proposal from the Church of England clergyman William Assheton, the Mercers' Company began a scheme in 1699 that would provide annuities to widows of clergymen and others in a similar class. It was intended that the scheme would be funded by subscriptions to it and only secured by alienating a small parcel of land. None of these private schemes were properly funded and many went broke. Within only four or five years of the 1692 Million Act, individuals were looking to buy or sell life annuities on the open market, often, but not always, secured by land. Between 1715 and 1720, the open market for life annuities was firmly established with at least one major broker in London in business in the mid-1720s. This was an obscure gentleman named Thomas Rogers.[22]

The growth of the life-annuity business is illustrated in the diagram that shows various kinds of annuity sales per quarter between 1690 and 1727. The bottom row shows the approximate times of floatation of various government annuity schemes to finance the growing national debt, whose beginning is traditionally traced to the Million Act of 1692. The row that is second from the bottom shows the approximate times that life annuity schemes available to the public were launched to finance some kind of undertaking, such as the Oil Annuity Office and the scheme from the Mercers' company.[23] The data for the top two rows were gleaned from late

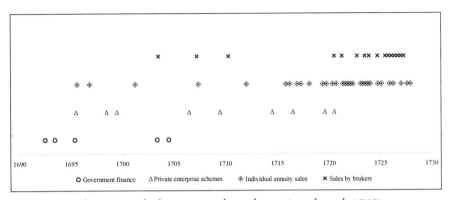

Life annuity sales by quarter and type from 1690 through 1727.

seventeenth- and early eighteenth-century English newspapers from the Burney Newspaper Collection in the British Library. Each symbol shows the quarter in which at least one newspaper advertisement is placed offering a life annuity for sale or showing someone's desire to purchase one. The top row of crosses is from advertisements placed by those who can be recognized as brokers; the second row of diamonds is based on what appear to be advertisements from individuals. What is not shown in these top two rows is the number of advertisements placed in each quarter. For individuals, the number is usually one, with a spike very soon after a stock market collapse called the South Sea Bubble in mid-1720 and another about three years later.

For brokers, there is a spike in 1723, again about three years after the South Sea Bubble. This spike is from a single broker named James Colebrooke, a London banker who advertised the sale of annuities on the lives of more than a dozen different people who are named in the advertisement.[24] The families had probably lost money in the market crash and needed to sell their government annuities. Beginning in 1724, the broker Thomas Rogers usually advertised three life annuities per quarter, either for sale or desired to be purchased, in among an often long list of properties for sale or lease.

Mathematicians writing commercial arithmetic books soon recognized the emerging market in life annuities. At least two responded by writing, or rewriting, their books to include material on life annuities. These were John Ward and Edward Hatton, both of London. Ward was initially a teacher of mathematics and Hatton was a surveyor, probably attached to one of the fire insurance companies in London.[25] Later, after a brief stint working in the Excise Office, Ward taught mathematics in Chester.

Ward was the first to mention Halley's work on life annuities. In the mid-1690s Ward wrote *A Compendium of Algebra*, which contains an appendix on interest and fixed-term annuities.[26] He took this appendix and expanded it into a chapter on the subject for *The Young Mathematician's Guide* published in 1707. Near the end of the chapter, Ward mentions Halley's 1693 paper, briefly summarizes the contents, and reproduces Halley's table of annuity values at the various ages of issue.[27] The same material appears in four later editions of *The Young Mathematician's Guide* that published before 1725. Halley's table of life annuities at 6% interest remained unchanged in the 1719 and 1724 editions, although the legal rate of interest had changed to 5% in 1714.[28] The amount of work required to redo the table of annuity values at a different rate of interest was probably prohibitive. Ward's synopsis of Halley's work is enlarged in his *Clavis Usurae; or a Key to Interest both Simple and Compound*, which was published in 1710.[29]

Edward Hatton published *An Index to Interest* in 1711 with a second edition in 1714. Hatton makes only a brief mention of the contents of Halley's paper and then delves right into life annuities.[30] Like Ward, he reproduces Halley's table of annuity values. But then he tries to do something new and different—he tries to approximate the values of annuities on two and three lives. Although incorrect, his approximation is an interesting one. For two lives, he takes the younger life, equates the value associated with a single-life annuity, taken from Halley's table, to a fixed-term annuity, and then solves for the term in that annuity. In the actuarial notation given earlier, for a chosen age x, Hatton sets $a_x = a_{\overline{m}}$ and solves for m, where a_x is taken from Halley's values. The approximation at this step is one that De Moivre and others later used. The problematical part in Hatton's valuation is in the next step. He takes the difference in the ages between the two lives and tacks the difference onto the term that he obtains in the first step. The resulting annuity with the newly derived term becomes the approximation to the joint-life annuity. Hatton republished his method in 1721 in a new publication, *An Intire System of Arithmetic*.[31]

There are some advertisements in the London newspapers for properties that had leases for lives. They are less prevalent than advertisements for life annuities. The earliest that I could find is from 1710.[32] The advertisement was made because an act of Parliament required that an estate in Devon be sold in order to pay a mortgage on it. Part of the estate comprised land leased on lives. Another advertisement six years later is more common among these uncommon advertisements.[33] A widow, Catherine Thompson, held the lease on lands in Hertford owned by St. Paul's, London. Typical of church lands, it was a lease for lives. In her will she stated that her legatee could sell the lease.[34] The sale was carried out under a decree of the High Court of Chancery and one of the masters in Chancery was in charge of the sale. In addition to selling lands held on leases for lives, the masters in Chancery also sold life annuities that were held by people who had been bankrupted.

This brings us back, finally, to Thomas Parker, 1st Earl of Macclesfield. As the person in charge of the Chancery Office, who was possibly getting a cut on the sales and other dealings by the masters in Chancery, he would have been interested

in what would be a good price for these life-contingent properties. In his capacity as chief judge in the High Court of Chancery, he also presided over cases that involved some complex property issues. For example, in 1724 a case came to Chancery in which a landowner died leaving debts.[35] His property, some of which was leased for lives and some for rack rents or yearly rental agreements, had been conveyed to his heirs, who were now responsible for the debts. The lenders wanted their money and Macclesfield, in his judgment, ruled that the properties with rack rents should be sold first to cover the debts. The properties leased on lives should be sold only if the former sale did not raise sufficient funds. This would simplify what to do with the fines that the heirs had already collected on the leases for lives. Macclesfield has also been described as "an acquisitive purchaser of land," so his interest in valuing leases for lives may have gone well beyond the courtroom.[36]

Given his connection to Macclesfield and other landed families, De Moivre's interest in life annuities was almost certainly motivated by financial issues related to land. As we shall see, the motivation comes out very subtly in his book. The ability to evaluate life annuities sold on the open market was an added bonus from which De Moivre benefited in later years.

De Moivre made a very simple insight into Halley's published mortality data. The insight can be obtained immediately by looking at a plot of the number alive at each age in the city of Breslau. After age 30, the curve is approximately linear, so De Moivre made the natural assumption that the survival probabilities after age 30 are linear in age. In order to formalize this model, denote the number of years left to the end of the table by n, i.e., let $n = \omega - x$. Then De Moivre's assumption on the survivor function can be expressed as

$$_tp_x = 1 - \frac{t}{n}.$$

This formula is also obtained from an equivalent assumption that a constant number of deaths occur at each age.

Using this linear survivor function, the value of the life annuity can be expressed as

$$a_x = \sum_{t=1}^{n} v^t - \sum_{t=1}^{n} \frac{tv^t}{n}.$$

The first term on the right is the sum of a geometric progression that is the present value of an annuity for a fixed term of n, or $a_{\overline{n}|}$. Since this value can also be expressed as $a_{\overline{n}|} = (1 - v^n)/i$, it can be easily calculated by hand with tables of logarithms. The second term on the right is an arithmetic-geometric progression. After a little algebra

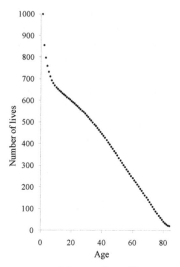

Halley's estimate of the number of lives at each age.

the term simplifies to

$$a_{\overline{n}|}\left(1+\frac{1}{n}+\frac{1}{in}\right)+\frac{1}{i}.$$

Combining the two terms yields

$$a_x=\frac{1}{i}\left(1-\frac{(1+i)a_{\overline{n}|}}{n}\right),$$

again an expression easily calculable by hand with tables of logarithms. Using some numerical examples, De Moivre showed that he could provide a good approximation to Halley's annuity values when he assumes $\omega = 86$ and $x \geq 30$. With this simple assumption, the calculation of the value of a life annuity, which for Halley was an onerous undertaking, now requires only a few calculations using logarithms. The saving in time is enormous. Using the same linear assumption, De Moivre also obtained an easily computable expression for the present value of an annuity paid for the life of the annuitant or a fixed term, whichever comes first.

Further, De Moivre showed how to carry out the evaluation of a_x when $_tp_x$ is piecewise linear. It took a little more effort in calculation but, in terms of Halley's life table, it was a realistic assumption when ages under 30 are included. From the plot, a reasonable assumption might be that the survivor function for Halley's data follows one type of linear trend for ages 15 to 30 and a different linear trend beyond age 30.

When it came to joint lives, De Moivre had to change gears. Using the linear assumption and the assumption that the lives are all independent, the present value of an annuity of 1 paid to three lives currently aged x, y, and z while they are all living is given by the formula

$$a_{xyz} = \sum_{t=1}^{\min(n,k,h)} v^t \left(1 - \frac{t}{n}\right)\left(1 - \frac{t}{k}\right)\left(1 - \frac{t}{h}\right),$$

where $n = \omega - x$, $k = \omega - y$, and $h = \omega - z$. Any simplification of this expression is very unwieldy, so De Moivre changed his assumption on survivorship to get an easily calculable result. He assumed that the survivor function has an exponential shape rather than a linear one and that, for convenience of finding sums, the potential length of life was limitless. For the survivor function, this yields $_tp_x = p'_x$. Then, for a single life under these assumptions,

$$a_x = \sum_{t=1}^{\infty} \left(v \cdot p_x\right)^t = \frac{v \cdot p_x}{1 - v \cdot p_x}.$$

This can be solved for p_x in terms of a_x to obtain

$$p_x = \left(1 + i\right) \frac{a_x}{1 + a_x}.$$

On applying the same principle, the joint-survivor annuity becomes

$$a_{xyz} = \sum_{t=1}^{\infty} \left(v \cdot p_x \cdot p_y \cdot p_z\right)^t = \frac{v \cdot p_x \cdot p_y \cdot p_z}{1 - v \cdot p_x \cdot p_y \cdot p_z}.$$

On substituting each of p_x, p_y, and p_z in terms of a_x, a_y, and a_z respectively, the formula

$$a_{xyz} = \frac{(1+i)^2 \cdot a_x \cdot a_y \cdot a_z}{\left(1 + a_x\right)\left(1 + a_y\right)\left(1 + a_z\right) - (1+i)^2 \cdot a_x \cdot a_y \cdot a_z}$$

is obtained. The value of each single-life annuity can then be determined using the linear assumption on the survivor distribution. Again, this is easily calculable by hand using logarithms.

De Moivre also considered the evaluation of last-survivor annuities. This is an annuity of 1 paid to a group of people until the last person expires. For three lives aged x, y, and z at issue, the present value of this annuity is denoted by $a_{\overline{xyz}}$ in the actuarial literature. Through simple probability arguments, it can be shown that

$$a_{\overline{xyz}} = a_x + a_y + a_x - a_{xy} - a_{xz} - a_{yz} + a_{xyz}.$$

The joint-survivor annuities can all be calculated using De Moivre's methods involving the mixture of linear and exponential survivor assumptions.

Using only one example, De Moivre checked the accuracy of making the incompatible linear and exponential assumptions together. For two lives, under the linear assumption only the present value of the last-survivor annuity is given by

$$a_{\overline{xy}} = \sum_{t=1}^{\min(n,k)} v^t \left(1 - \frac{t}{n}\frac{t}{k}\right).$$

De Moivre was able to evaluate this directly and compare it to the value of $a_{\overline{xy}} = a_x + a_y - a_{xy}$, where a_{xy} is evaluated using the mixture of the linear and exponential survivor assumptions. For $x = 36$ and $y = 46$, he found, incorrectly,[37] very close agreement between the two methods of calculation. After that, he never questioned the incompatibility of the two assumptions.

From these basic results, using the linear assumption, possibly in combination with the exponential assumption, much of the remainder of De Moivre's results follow; only the situation changes. De Moivre considered three additional situations with respect to survivorship: (1) reversions, (2) successive lives, and (3) renewal of lives. The first two of these situations pertain both to life annuities bought and sold on the open market and to annuities related to land tenure. The third one pertains only to land—in particular, to leases for lives. In the first type of survivorship, a reversionary annuity for a given person is one that makes lifetime payments to that person beginning on the death of another person. There are variations on this annuity. It could be paid to one person after the death of the last survivor of two other people. It could be paid on the joint lives of two people beginning on the death of a third. In a succession of lives, the annuity is initially paid to a given annuitant. On that person's death, the payment is made to a successor named by the original annuitant. When there is an option for the renewal of lives, payments could continue indefinitely. Without renewal, payments would be made until the last survivor dies. If the annuitants exercise their option of renewal, then on the death of one of the annuitants, an amount is paid (a fine) to replace that person with someone else. The renewal of lives only occurrs in the case when there was a lease for lives.

Reversionary annuities were available on the open market; their origins, however, were in land tenure and the protection of widows and children of the landlord.
The tradition of the marriage settlement is one typical example. Here, a husband
contracts, prior to marriage, that on his death his wife will receive annual payments
for her life. Payments to the widow are guaranteed by the husband alienating part
of his land for this purpose. The Mercers' Company annuity scheme is a type of
reversionary annuity that was available on the open market. By paying into a fund,
payments could be made to the widow after the death of her husband. This would be
attractive to someone like a clergyman whose income derived from land. This was
not land that he could alienate, but rather land owned by the Church of England. On
the clergyman's death, his income from the Church ceased and so payments to the
family would also cease.

At this point, it is useful to provide an annuity problem that William Jones set
for his students around 1740.[38] The problem is to find the fine for a renewal of a lease
for lives. As stated by Jones, the problem runs,

> An estate of 700£ p ann consisting of 14 Farms of 50£ p. ann each. is let on
> leases of three lives: and upon the failure of any one of the three lives in each
> lease, the Landlord is to receive a fine of £5; and to add any new life to the lease
> that the Tenant pleases to name. Quere how much yearly rent ought to be added
> to the Rent Roll of the said Estate, on account of the chances of Fines?

The question is actually based on a request that Jones received from the Scottish
aristocrat John Hope, 2nd Earl of Hopetoun, in 1740. Hopetoun's letter opens with

> In this country leases for lives being a new thing[;] we are not much acquainted
> with the method of valuing them, and it happens at present to be of some
> consequence to me to have a true solution of the following question which I
> know no body can give me with more accuracy than you.

What follows immediately in Hopetoun's letter is exactly the same question that
Jones gave to his students. The connection of annuity valuations to land tenure is
explicit. And the involvement of a mathematician as consultant is interesting.

Unlike Jones's correspondence with Hopetoun, the connection of De Moivre's
Annuities upon Lives to land tenure is not explicit in the book. This is a source of
confusion to many when trying to connect the book to applications. Without any
explicit connection, the seemingly most obvious possibility for the modern reader
is the open market in life annuities, as the book begins with the evaluation of life
annuities. However, the section in the book on successive lives and reversions,[39] like
Jones's correspondence, points directly to the practice of copyhold leases and not to
the usual annuity market. For example, De Moivre's Problem 16 in *Annuities upon
Lives* is a direct answer to the problem of valuing a lease on three lives when there

is a fine to be paid on the replacement of any expired life. The problem reads, "If there be three equal Lives, and A or his Heirs are to have a Sum f paid them upon the Vacancy of any of those Lives, what is the Expectation of A worth in present Money." De Moivre's method of solution is based on the exponential survivorship model. In an undated letter to Jones, De Moivre solved a simpler version of this problem the same way and referred to the sum paid specifically as a *fine*,[40] the usual term in the renewal of a lease for lives.

De Moivre's *Annuities upon Lives* had impact very soon after its publication. It occurred in three ways: (1) his results resonated among his friends and connections, one of whom was highly placed politically; (2) more publications, directly inspired by De Moivre's book, appeared on the valuation of life annuities; and (3) some mathematicians, as well as others, began to offer their services to the public as consultants in the valuation of life annuities. The writers of annuity books were typically not necessarily professional, or even accomplished, mathematicians. On the other hand, they did have an intimate knowledge of either the London financial industry or of landed estates. We have already seen Jones working as a consultant. De Moivre also worked as a consultant, holding office at Slaughter's Coffeehouse.

Newton owned a copy of *Annuities upon Lives*.[41] So did Halley.[42] William Jones may have owned a copy. He certainly had access to one; he made an extensive set of notes on the book, notes which have survived.[43] Perhaps not satisfied with De Moivre's linear assumption on the survivor curve, Jones tried to fill in the missing ages at the end of Halley's Breslau table. In that way, annuity values could be computed based on a complete table, unlike De Moivre's model. Brook Taylor also filled in missing ages in Halley's table and then went on to calculate, by hand, the present value of a life annuity at each issue age from 1 through 90 using interest rates of 4% and 5%, a very time-consuming task. Based on these single-life annuities, he also calculated the present value of some life annuities for two, three, and four lives.[44] In each case, he used De Moivre's method to evaluate joint-life annuities. It comes as no surprise that Macclesfield also possessed a copy of *Annuities upon Lives*.[45] Although a highly placed politician, he was not the one that made calculations, or had the calculations made for him, based on De Moivre's methods. Among the papers of Robert Walpole is a set of annuity valuations.[46] There is a table whose entries contain the present values of life annuities at issue ages 1 through 68 using an interest rate of 4%. The claim at the head of these valuations is that they were based on Halley's Breslau table. Where De Moivre enters is that there are valuations for a few joint-life annuities for two and three lives. Like Brook Taylor's, these valuations use De Moivre's methodology for joint-life annuities. Other than William Jones's manuscripts, there is no surviving evidence to indicate whether any of them applied any of the results that De Moivre obtained in the book to the actual purchase of any form of life annuities.

In 1727, a book entitled *A New Method for Valuing of Annuities upon Lives*[47] appeared by the London accountant and writing master Richard Hayes. He wanted to fill a niche left empty by De Moivre. *Annuities upon Lives* requires some mathematical facility in order to come up with an actual numerical valuation of a life annuity. Consequently, Hayes's book contains an extensive set of tables setting out the present value of a life annuity for a single life at several possible values of annual payments (£1 to £10 in steps of 1, £10 to £100 in steps of 10, and £100 to £1000 in steps of 100) each at rates of interest 4%, 5%, 6%, 7%, and 8%. The age at issue for the annuitant ranges from 30 to 73. Prior to this book, Hayes had written several other books related to various financial transactions in the city.[48] Together, Hayes's books indicate an intimate knowledge of a wide range of the mercantile and financial sector in London. Hayes must have thought the new tables for life annuities would be useful for traders in the market.

After Hayes, several books on life annuities came out in the 1730s. They were all aimed at landlords. In his 1730 book, *The Gentleman's Steward and Tenants of Manors Instructed*,[49] John Richards of Exeter describes in detail the different methods of land tenure. He gives many numerical examples of how to calculate the value of these tenures, taking into account various types of rent and other fees attached to the estate. In all of these calculations, he takes into account the probabilities of life. Richards uses De Moivre's linear assumption for the survivor distribution and provides tables at 4%, 5%, 6%, 7%, and 8% for the values of single- and joint-life annuities at various ages.[50] Richards's book was soon followed by three others: an Irish edition of De Moivre's *Annuities upon Lives* published in 1731 and two books on evaluating annuities and leases, one by Gael Morris published in 1735 and the other by Weyman Lee published in 1737.[51] What these books have in common is that they all mention De Moivre's work on annuities, though some do not use his results but try to come up with their own methodology, including use of the approximation $a_x = a_{\overline{m}|}$. Macclesfield owned a copy of Richards's book. After his death in 1732, his son obtained copies of Morris's and Lee's books.[52]

The Irish reprint of *Annuities upon Lives* was probably not sanctioned by De Moivre. One reason that the book appeared in print in Ireland was that the Irish printer involved saw a market for the book and at the same time did not have to pay any copyright fees in England.[53] As to the market, leases for lives were common in Ireland. As a result of Irish Catholic support for James II during the Revolution of 1688 and later, laws were enacted that restricted land ownership by Catholics in Ireland. The landowner, usually Protestant and sometimes an absentee, let his land to a number of prosperous middlemen, often Catholic, who sublet the land to tenant farmers.[54] The lease for the middleman was equivalent to purchasing an annuity, where the annual payments were the rents collected from the tenant farmers above the middleman's own rent. The fine was equivalent to the sum required to purchase this annuity. The material added to De Moivre's original is concerned initially with life-annuity valuations (using the relationship $a_x = a_{\overline{m}|}$ that Hatton had used earlier) and then with issues related to the purchase of freeholds and the present value of fines for lease renewals.[55]

These books did not, as Loraine Daston has pointed out, have much impact on setting the price of annuities as they were offered in the market. They were, I believe, used as reference guides for buyers. In addition to the earls of Macclesfield owning these books, two examples support this belief, one from real life and one from fiction.

Henry Stewart Stevens was a barrister and one-time student of De Moivre. By the 1740s, Stevens had purchased property in the countryside and had set himself up as a country squire in Berkshire. Soon tiring of country life, Stevens wanted to purchase a life annuity on his own life, thus having an annual income without the worry of estate management. With this and other money he had, he planned to leave his squiredom and settle in France or Italy.[56] Someone else, a physician named Richard Allen, had already purchased an annuity from the Exchequer Office on Stevens's life that paid £100 per year. Stevens wanted to buy this annuity from Allen. On February 10, 1744, Stevens wrote to Allen with an offer. Receiving no reply, he then wrote to one of Allen's neighbors outlining the reasons why Allen should sell; essentially, Allen needed a large amount of cash to settle an estate. Stevens offered one thousand guineas (£1050) for the annuity. He argued that the amount was a good price since he felt that if Allen sold the annuity on the open market he would get only £700 for it,[57] the standard seven-years purchase. Stevens took his offer price from Thomas Simpson's 1742 *Doctrine of Annuities and Reversions*,[58] the most current, mathematically sound pricing he could find. He even included a copy of the book in his correspondence with Allen's neighbor. Allen must not have sold. Stevens remained a country squire and eventually held the position of deputy lieutenant of the county of Berkshire.

The fictional example is from Henry Fielding's *Tom Jones*, published in 1749. A basic outline of the story is that Squire Allworthy is rich and has no children. His sister Bridget has married a Captain Blifil and they have a son, the odious Master Blifil who is to be Allworthy's heir. The Captain had married Bridget for her money. There is also an illegitimate child, Tom Jones, whom Squire Allworthy has taken in and raised with Master Blifil. In Chapter VIII, Book II of the novel, Fielding writes of Captain Blifil's plan to use Allworthy's money to make changes to the house and gardens of the estate on Allworthy's death:

> Nothing was wanting to enable him to enter upon the immediate Execution of this Plan, but the Death of Mr. Allworthy; in calculating which he had employed much of his own Algebra besides purchasing every book extant that treats of the Value of Lives, Reversions, &c. From all which he satisfied himself, that as he had every Day a Chance of this happening, so had he more than an even Chance of its happening within a few Years.[59]

Of course, Captain Blifil and Bridget die before Allworthy. After some adventures surrounding Tom's love interest, Sophia Weston, who is engaged against her wishes to Master Blifil, it turns out that the good-hearted but illegitimate Tom is Bridget's eldest son. With Master Blifil revealed to Allworthy as the odious man he is, Tom becomes Allworthy's heir, marries Sophia, and all ends happily. Although his actions are the subject of satire by Fielding, the fact that Captain Blifil had his head buried in some annuity books indicates that reading these books was a fairly common activity among the landed, as it was for the Macclesfields, apparently.

Beyond the books, another reference guide for the annuity or lease purchaser was the private consultant. There is very little surviving evidence of this activity in and around the 1730s, but it does exist. William Jones's papers provide some examples. In addition to his 1740 correspondence with the Earl of Hopetoun, Jones also handled two other queries in the 1730s. On May 12, 1732, fellow Welshman Moses Williams wrote to Jones asking him for help. A woman aged 32 was receiving an annuity of £200 for her lifetime and the lives of two others, a man aged 65 and his wife, aged 63. On the death of the man and his wife, the woman was to receive £400 per annum. Williams asked Jones to determine for him the value of the annuity and the value of the reversion. A year later, John Fortescue wrote to Jones with an annuity question. Fortescue was to pay a man a lump sum of £4500. The man was willing to take the money as a life annuity and so Fortescue wanted to know what the annual payment would be when the interest rate was 3 or 3.5%.[60]

There were others, besides Jones, who did this kind of consulting. In a 1729 advertisement directed at the nobility and gentry, the land surveyors John and Samuel Warner indicated several services they offered in addition to their normal surveying and mapping of estates:

> They [the Warners] likewise Compute Interest, Estimate the Value of Annuities for Lives, Calculate the present Worth of Leases, Rents or Pensions in Possession or Reversion at any Rate of Interest, Simple or Compound.[61]

If they were advertising, they must have perceived that there was a market for life-contingent calculations, however crudely done, related to land.

After the publication of *Annuities upon Lives*, Abraham De Moivre also carried out work as an annuity consultant in addition to, and possibly eventually in place of, his work as tutor. He also, to a lesser extent, gave advice on games of chance.[62] There is only one known annuity valuation in De Moivre's hand.[63] It is undated, through in all probability it is post-1725.[64] Also, the recipient of the valuation is unknown and there is not enough information to determine how De Moivre obtained his numerical results. All that is known is that De Moivre evaluated an estate in which a reversion was involved. The valuation was from the point of view of someone buying the reversion; De Moivre was providing guidance to the buyer. The estate was originally worth "25 years purchase." De Moivre gave his evaluation for three different scenarios depending on how many years had elapsed since the estate was first purchased.

When De Moivre published the second edition of *Doctrine of Chances* in 1738, he placed an advertisement in a London newspaper in July of the next year.

Initially, it was addressed to the subscribers of the book who had not obtained their copies. After telling his subscribers that they could pick up their book at Slaughter's Coffeehouse, De Moivre added,

> He also takes this Opportunity of making it his Request to those, who are pleased to consult him by Letter about any Case relating to Leases, for a Number of Years certain, or to Annuities upon Lives, to mention what the Rate of Interest is, which is agreed upon between the contracting Parties.[65]

De Moivre dealt with his clients in confidence regarding these valuations. Presumably, his clients were bargaining with others and came to De Moivre to seek a reasonable price without the other side's knowledge.[66] The confidential nature of his dealings with clients is seen in an advertisement he placed in another London newspaper two years later in 1741.

> If the Gentleman who Yesterday spoke to Mr. A. De Moivre upon the Pavement in St. Martin's-Lane, about a Question relating to Annuities on Lives, will come to Slaughter's Coffee-house, he shall receive Satisfaction about the same; and his Pardon asked for the Answer he received, it being not immediately recollected, by Reason of different Dress and other Circumstances, that he was the same Person who some Days before had proposed the Question.[67]

Once more De Moivre's cantankerous nature begins to rise up, this time while he is trying to maintain the confidentiality of his clients. One can only imagine the interaction between De Moivre and the gentleman on the pavement as De Moivre thought the person in front of him was trying to get information on someone else's business dealings.

There is a hint of De Moivre's recognition of the necessity to keep a confidence that appears in an autograph book that he signed for the Church of England clergyman and antiquary Cox Macro.[68] De Moivre and Macro may have met through mutual connections with Joseph Goupy, designer of the frontispiece for *Doctrine of Chances*, or perhaps Goupy's pupil in drawing, Brook Taylor. Macro was Goupy's patron.[69] Macro's book contains entries from several people; each person provided a signature with some kind of quotation or, in one case, a drawing. De Moivre wrote in the book, "Est et fideli tuta silentio merces" and then signed his named. This is a line taken from Book III.2 of the *Odes*, written by the lyric poet Horace of ancient Rome. A translation of the line is "There is also sure recompense for faithful silence." The poem originally advised the reader not to divulge the mysteries of the gods;[70] but "modernized" to the eighteenth century, it equally applies to De Moivre's relationship with the gentleman on the pavement outside Slaughter's Coffeehouse.

A smidgeon of evidence survives indicating that De Moivre's methods for the valuation of life-contingent contracts related to real property were widely used.

Just prior to De Moivre's death in 1754, but well after the 1st Earl of Macclesfield's death, the High Court of Chancery explicitly recognized De Moivre's work. The recognition came in the form of a rejection. In a case that it considered involving an undervalued reversion on an estate, the court ruled that it could not impose De Moivre's methodology.[71] The key issue for the court was that there was no fraud involved when the two parties came to an agreement on the purchase price. The court did recognize the good deal that the purchaser of the reversion enjoyed, and so did not award any court costs to him after it ruled in his favor.[72]

The foothold established by annuities as an application of probability moved further from land and more into the general life annuity marketplace as the eighteenth century progressed. By the end of the eighteenth century, attention to annuity and related calculations became the main focus of British activity in probability.

❧ 12 ❧
The Decade of the Doctrine Enhanced

De Moivre's *Miscellanea Analytica* appeared in 1730.[1] The only known advertisement for its availability on publication is in the May 1730 issue of *The Present State of the Republic of Letters*.[2] The book was sold by subscription with De Moivre controlling the distribution of the copies. This situation is the same as the first edition of *Doctrine of Chances*. The difference is that the subscription list to *Miscellanea Analytica* was published with the book.

A careful examination of the list provides some insight into De Moivre's knowledge community: his close friends, his colleagues, his students, and his patrons. The great majority of subscribers were a small group of the Whig political elite that had close ties to one another through blood, through marriage and through the vast political patronage web set up by Robert Walpole that was at the foundation of how he ran the government. There were a few Huguenot friends and most of his surviving mathematical friends: John Colson, Alexander Cuming, Martin Folkes, William Jones, John Machin, Colin Maclaurin, Edward Montagu, Nicholas Saunderson, James Stirling, and Brook Taylor. Newton was dead by this time, but he was represented by John Conduitt, who had married Newton's niece. For some unknown reason, his long-time friend and colleague Edmond Halley is missing from the list. Beyond the mathematicians, some members of the Royal Society were among the subscribers. Many of this group were physicians, aristocrats, or Members of Parliament.[3]

The choice of publishers for *Miscellanea Analytica*, Jacob Tonson and John Watts, also reflects De Moivre's high-level Whig connections. There are two Jacob Tonsons—uncle and nephew. The nephew was the active publisher in 1730 and therefore responsible for getting *Miscellanea Analytica* into print. The uncle had retired from working as a publisher in about 1720, but remained active in editing and in advising his nephew. The uncle was also the one with strong political connections. He was a founder and permanent secretary of the Kit-Cat Club, which was devoted to furthering Whig political objectives. Its membership included leading literary figures and politicians. Robert Walpole was a member. About one-fifth of the known members of the club, or their surviving sons in 1728, subscribed to *Miscellanea Analytica*.[4] With the exception of Walpole, these subscribers from the Kit-Cat Club were all aristocrats. It is likely that De Moivre's political connections helped him to secure the Tonsons as the publishers for his book.

Most subscribers to *Miscellanea Analytica* bought a single copy, perhaps for themselves or perhaps just to support De Moivre. Others supported De Moivre more lavishly. John Conduitt purchased fifteen copies and former pupil John Montagu, 2nd Duke of Montagu, purchased ten. They may each have kept one, if that. Other multiple-copy subscribers had an interest in mathematics as well as in supporting De Moivre. Martin Folkes, to whom *Miscellanea Analytica* is dedicated, purchased seven copies; three remained in his library at his death, including a presentation copy printed on large pages with gilded edges.[5] The presentation copy was bound in Turkish leather, imported goatskin that was used only in the highest quality bookbinding work.[6] Colin Maclaurin purchased six copies, all meant for mathematical colleagues in Scotland.[7]

By examining the subscription list in a slightly different way, attempts can be made at answering two questions that come to mind: Why did De Moivre put this material into a book rather than writing a number of articles that would appear in *Philosophical Transactions*, for example? Why did some wealthy individuals, aristocrats among them, buy *Miscellanea Analytica* when they may not have been interested in the actual contents of the book?

A simple answer to the first question might be found by ignoring the subscription list and examining the Royal Society's treatment of mathematics, both at its meetings and in its publications. A read through the Royal Society's minutes of their meetings in the journal books shows scant attention paid to De Moivre's primary research interests—topics in pure mathematics.[8] An examination of *Philosophical Transactions* shows a decline in interest in mathematics between the two decades of the 1710s and the 1720s. From 1710 to 1719, about 10.5% of the papers were on topics related to pure mathematics or mathematical physics. From 1720 to 1729 the percentage declined to about 2.5%. The simple, but incomplete, answer is that De Moivre needed an outlet for his mathematical publications, and with declining interest in mathematics within the Royal Society, he had to create his own. Due to the continuing necessity to teach to make a living and the time that it involved, he accumulated his work over several years and then went to his Whig patrons to get his work into print.

There is a more complex answer to the first question. In the introduction to his work that describes the priority dispute between Newton and Leibniz, Rupert Hall outlines some of the differences between the pursuit of mathematics today and in the eighteenth century. After noting that "success in the scholarly or academic world depended far more on a militant combativeness then than it does now," Hall goes on to say:

> To put it crudely, an achievement in scholarship, science, mathematics, or medicine was a marketable commodity, a highly personal property: The recognition it conferred might be a first step toward attainment of a bishopric or an office of state. And the rules of the marketplace were both capricious and very different from those that now prevail. From the late nineteenth century, peer evaluation has been the rule of science and learning in the civilized world; and laymen have largely accepted the judgment of the internal experts. In the lifetimes of Newton and Leibniz what counted most was not the opinion of one's peers but the direct impression made on princes and ministers, prelates and magnates, who exercised enormous personal powers of appointment.[9]

De Moivre had a number of ministers and magnates on his subscription list. As late as 1730, he may have been continuing to demonstrate his talents in pursuit of the elusive patronage position that never came his way.

To try to answer the second question, it is useful to think of *Miscellanea Analytica* as a luxury item in the book world. One guinea was near the high end as a price for a book. By purchasing one or more subscriptions, and regardless of whether the book was taken home or left with De Moivre, the subscriber could be identified as a patron of the leading mathematician in England after Newton. Having the book on the subscriber's bookshelf would be an indication of that patron's erudition or intellectual union with an important aspect of science. In a sense, the subscriber identified himself with the scientific quality of the author. Horace Walpole claimed to be very poor at mathematics,[10] and yet had a copy of the first edition of De Moivre's *Doctrine of Chances* in his library.[11] It would not have been an easy read for Walpole, but its display would indicate Walpole's awareness of the importance and popularity of the subject, as well as the importance of the book's author.

For De Moivre, the sale of *Miscellanea Analytica* by subscription had two advantages. The first is that he knew he could cover the printing costs prior to publication. The second, and more important, reason is that, since he controlled the distribution of the book, he could give copies of the book to those nonsubscribers who he thought should receive it. These gift copies would be taken from any extra copies De Moivre had printed or from the books left with him by subscribers. This was probably how James Stirling received his copy of the first edition of *Doctrine of Chances*. For *Miscellanea Analytica*, this type of distribution is illustrated by the situations of Nicolaus and Johann Bernoulli. Despite the fact that they had not corresponded with De Moivre for fifteen years or more, both Bernoullis likely received copies of *Miscellanea Analytica* directly from De Moivre. In a 1731 letter to Nicolaus Bernoulli, Gabriel Cramer wrote that if Nicolaus had not yet received a copy of *Miscellanea Analytica*, he should soon receive one from De Moivre. The reason was that during Cramer's visit to London prior to the publication of *Miscellanea Analytica*, De Moivre told Cramer of his plans to send copies of his book to both Johann and Nicolaus Bernoulli.[12]

As its title suggests, *Miscellanea Analytica* is a collection of results on a variety of mathematical subjects. Several of the major topics, whose results date from the mid-1720s and before, were described in Chapters 7 and 9. Material in *Miscellanea Analytica* that was discussed in Chapter 7 includes De Moivre's unique approach to summing infinite series of numbers and his work on centripetal forces. Two topics described in Chapter 9 are the so-called "De Moivre's theorem," which states a trigonometric relationship involving $i = \sqrt{-1}$, as well as the theorem's application to solving Cotes's problem regarding trinomial divisors; and the development of a generating function to solve a problem in probability dealing with the sum on the faces that show in the throw of several dice. There is much more material that appears in *Miscellanea Analytica*. De Moivre provides greater detail to his trigonometric solution to the duration of play problem, as well as a further treatment of recurrent series. He delves into enough additional problems in infinite series that the French academician, Pierre de Maupertuis, who was not interested in these types of problems but subscribed to the book anyway, found *Miscellanea Analytica* not to his liking. Maupertuis' subscription was probably connected to De Moivre sponsoring Maupertuis for fellowship in the Royal Society during the academician's visit to London in 1728. After much ink spilt on series, there is an entire section in *Miscellanea Analytica* devoted to responding to Montmort, although the man had been dead for a decade or more. Finally, there is his major contribution to probability that treats the normal approximation to binomial probabilities.

The part of *Miscellanea Analytica* containing the response to Montmort (Book VII) has the title "Responsio ad quasdam criminationes" or "Response to certain accusations," in translation. De Moivre begins his response by giving a brief outline of how he began to work in probability theory and of Montmort's response to reading *De Mensura Sortis* given in his letter to Nicolaus Bernoulli that appears in the second edition of *Essay d'analyse*. This is followed by a list containing several of Montmort's comments and complaints about De Moivre's work. To end this introductory material, De Moivre gives a general defense of his own work. The particulars take up the rest of Book VII. In the middle of his defense, De Moivre could not resist quoting part of Bernard de Fontenelle's éloge of Montmort from 1719. In the quoted extract, after mentioning De Moivre in the éloge and the dispute with Montmort, Fontenelle did a double entendre on Montmort as lord of the manor in the role of landowner and mathematician. De Moivre showed that he understood the double entendre by writing in *Miscellanea Analytica* that he asked Pierre Varignon to pass on his thanks to Fontenelle for the compliment he received in Montmort's éloge.

Book VII shows De Moivre at his best and at his worst. He is at his intellectual best in the development of his generating function to solve the dicing problem for the sum of the faces that show in the throw of the dice. This is clouded somewhat by a preface containing a long discussion devoted to why he could not possibly have relied on either edition of Montmort's *Essay d'analyse* to obtain his answer, which appeared without proof in both *De Mensura Sortis* and *Doctrine of Chances*. There is no mention in *Miscellanea Analytica* that his solution using a generating function was obtained after the publication of *Doctrine of Chances*. All the manuscript evidence points to a different method of solution prior to the 1720s that was independent of Montmort's approach, with the generating function dating from the mid-1720s. Further into the dark side is De Moivre's explanation of why he and Montmort stopped corresponding. He claims that both he and Montmort were too busy. I have already commented on this in Chapter 7; as I said, De Moivre's explanation just does not ring true.

There is a lot more to Book VII. For example, Montmort had criticized De Moivre for his use of finite differences without referencing Montmort's work in *Essay d'analyse* when solving some problems in probability that require summation of series. De Moivre's first technical response in Book VII to Montmort's complaints is to review the theory of finite differences from Newton's original work and then to build on it substantially. Using one of Montmort's problems in finding the sum of a particular infinite series, De Moivre shows that there is more than one way to obtain the solution, the method of finite differences being only one of them.

After finite differences, De Moivre goes on to treat combinatorial problems in Book VII. Initially, he notes favorably that Montmort had developed methods for finding combinations based on Blaise Pascal's arithmetical triangle from the middle of the seventeenth century and on methods put forward by John Wallis later in the seventeenth century. De Moivre then writes that Montmort's treatment of combinations went beyond what had been done earlier. On the other hand, De Moivre found that for some problems it was difficult to apply the usual methods, so an approach using probability was preferable.

He begins the discussion of his probability approach by illustrating it with simple permutations and combinations. Suppose, for example, that C is the number of combinations of n different objects taken r at a time. By De Moivre's definition of probability, the probability of obtaining a particular combination is $1/C$. Consequently, if we can find the probability of a particular combination occurring, by inverting the solution we can find the number of combinations. For example, consider finding the probability of selecting three different items from n in any order. De Moivre would argue this as follows: the probability of selecting any of the three items is $3/n$; given that one item has been selected, the probability of selecting one of the other two is $2/(n-1)$; and given the first two selections, the probability of obtaining the last item from those that remain is $1/(n-2)$. The required probability is the product $(3/n)(2/(n-1))(1/(n-2))$, so the required expression for the number of combinations is $n(n-1)(n-2)/(3 \cdot 2 \cdot 1)$. After simple examples such as this, De Moivre went on to consider much more complicated ones.

What might be considered a quirky approach to combinatorics found some traction in Britain. Thomas Simpson acknowledged this approach in 1740 at the very beginning of his *Nature and Laws of Chance* as he was dealing with simple permutations and combinations.[13] Much further removed is the influence on the Reverend Philip Doddridge. Until his death in 1751, Doddridge used De Moivre's method to justify the derivation of simple permutations and combinations in his mathematics lectures at the dissenting academy at Northampton.[14] The way in which it is set out in his lecture notes indicates that Doddridge used the method to teach his students how to compose and write down a good logical argument.

Some of Book VII contains new material such as the development of his generating function for the dicing problem. Other parts of Book VII are a rehash of old material with some small additions and comments. For example, De Moivre goes over Woodcock's problem that he solved and published in *Doctrine of Chances*. The problem is a variation on the gambler's ruin problem with the pot increasing in a specified way after each round of play. In his treatment of Woodcock's problem in *Miscellanea Analytica*, De Moivre initially stated the solution that he had obtained in *Doctrine of Chances* in 1718.[15] Then he quoted an extract from a letter written by Nicolaus Bernoulli to De Moivre in which Bernoulli gave a simple solution to the problem. The extract had also appeared earlier in *Doctrine of Chances*, followed by an acknowledgment from De Moivre that Bernoulli's solution was very simple. Following the extract from Bernoulli's letter in *Miscellanea Analytica*, De Moivre played a little tit-for-tat with Bernoulli by giving an even simpler solution to Woodcock's problem. He said that his former student, the barrister Henry Stewart Stevens, had shown him the solution in 1720 or 1721. Then he said that after Gabriel Cramer visited Nicolaus Bernoulli in the late 1720s, Bernoulli wrote to Cramer informing him of a new solution to Woodcock's problem. Cramer passed the news on to De Moivre when he visited him in London in 1728. De Moivre did not see Bernoulli's new solution. Without checking with Bernoulli—De Moivre had ceased corresponding with him over a decade before—De Moivre guessed that Bernoulli's solution was the same as the one from Stevens. He staked out a priority claim without having any information of a rival claim.

Late in 1721, Alexander Cuming suggested a probability problem for De Moivre to work on. De Moivre claimed that he found a solution the next day.[16] The two probably met through the Royal Society; Cuming was elected a fellow in 1720. Cuming is an interesting character who had a checkered career.[17] He became a baronet on the death of his father in 1725. The height of his career occurred in 1730 when he travelled to America and returned with a number of Cherokees to sign a treaty between Britain and the Cherokee nation. Seven years later he was in debtors' prison, where he remained for eighteen years. He was still throwing out challenge problems in probability from debtors' prison in the 1740s.[18] He eventually died in poverty. His 1721 challenge problem to De Moivre resulted in one of De Moivre's major accomplishments in probability—the normal approximation to binomial probabilities.[19]

The whole of Book V of *Miscellanea Analytica* is about binomial probabilities, expressed in terms of the expansion of $(a + b)^n$, where a is the number of favorable chances and b is the number of unfavorable ones. The exponent n is the number of experiments carried out, trials made, or games played. The probability of success, or of achieving a favorable outcome, is $p = a/(a + b)$.

Cuming enters in Chapter II of Book V. It is not clear exactly what it was that Cuming suggested to De Moivre to work on and what were De Moivre's additions and extensions to the problem. De Moivre broke this section of *Miscellanea Analytica* into four problems. The first two problems deal with the deviations from the binomial mean. The third problem is about finding an approximation, when n is large, to the middle term in the expansion of $(1 + 1)^n$ all divided by 2^n. In terms of outcomes of a binomial experiment, the middle term is associated with $n/2$ successes and $n/2$ failures when n is an even number. The fourth problem that De Moivre considers is how to relate other binomial probabilities to the middle term in the expansion.

The first problem has a very simple statement. Two players of equal skill, A and B, play n games. At the end of these games, the player who wins the majority of games gives a spectator a number of units of money corresponding to the difference between the number of games the player has won and $n/2$. What is the expected amount of money that the spectator is to receive? In modern probability notation, let X be the number of games that player A wins. The random variable X has a binomial distribution with the number of trials equal to n and probability of success $p = 1/2$. What is required is to find $E(|X - n/2|)$, where E denotes the operator to find an expected value. De Moivre showed that this expectation is $n/2$ times the middle term in the expansion of $(1 + 1)^n$, all divided by 2^n. Written in its ugliest detail, the solution is

$$\frac{n}{2} \times \frac{n(n-1)(n-2)\cdots\left(n-\frac{n}{2}+1\right)}{\left(\frac{n}{2}\right)\left(\frac{n}{2}-1\right)\left(\frac{n}{2}-2\right)\cdots 3\cdot 2\cdot 1} \times \frac{1}{2^n}.$$

De Moivre followed this problem by a second one that generalizes the first. The players A and B no longer have equal skills. Instead, their respective skills are given by p and q, where $p + q = 1$. In this case, it is required to find $E(|X - np|)$, where it is assumed that np takes on an integer value, call it m. De Moivre shows that his expectation in this case is given by

$$\frac{2m(n-m)}{n}\binom{n}{m}p^m q^{n-m},$$

which reduces to the ugly expression of the answer to the first problem when $p = 1/2$ and $m = n/2$. As before in Chapter 5, the combinatorial coefficient is given by

$$\binom{n}{m} = \frac{n!}{m!(n-m)!}.$$

Since the answers to these first two problems are not obvious and consequently the middle term of the binomial did not initially arise, it may be that these two problems were Cuming's initial suggestions and that this was what De Moivre solved in a day. Even if this is the case, the remaining problems were solved soon thereafter since De Moivre in 1733 dated his discovery of the solution to the third problem to 1721.

De Moivre's third problem in Chapter II of Book V follows naturally from the first. Since the solution to the first problem involves the middle term in the expansion of $(1 + 1)^n$ divided by 2^n, it is natural to consider an approximation to this expression when n is large. Large n was a situation that Johann and Nicolaus Bernoulli previously considered. The approximation involves infinite series expansions,[20] thus supporting De Moivre's dictum that all problems in probability could be solved through the use of the binomial theorem or by infinite series, in this case together. In *Miscellanea Analytica*, De Moivre gives

$$\frac{2\frac{21}{125}\left(1-\frac{1}{n}\right)^n}{\sqrt{n-1}}$$

as the approximation for large n.

Given the middle term, which is also the term of maximal value in this case, it is also natural to consider other terms in the binomial expansion. This is the fourth problem that De Moivre considers in Chapter II of Book V. With the middle term associated with $m = n/2$ successes, De Moivre attacked the problem of finding the term that has $m + l$ successes. Denoting the value of the maximal term by M and the term with $m + l$ successes by Q, De Moivre obtained the ratio

$$\frac{M}{Q} = \frac{(m+l-1)^{m+l-\frac{1}{2}} \times (m-l+1)^{m-l+\frac{1}{2}} \times \frac{m+l}{m}}{m^{2m}}$$

to express the relationship between the two terms.

The remainder of Book V, Chapters III and IV, deals with other aspects of the binomial expansion, or distribution as it applies to probability. In Chapter III, De

Moivre derives the maximal term in the expansion of $(a+b)^n$. His result depends on issues such as whether n or $n-1$ can be expressed as integer multiples of $a+b$. In Chapter IV, De Moivre derives the points of inflection of the binomial.

The first two problems in Chapter II, or perhaps all of Chapter II, look like they were motivated by a result in Jacob Bernoulli's *Ars Conjectandi* that is the first expression of the law of large numbers. Rather than the expectation of $|X - np|$, Bernoulli was interested in probabilities concerning $|X/n - p|$. He wanted to find n, or k since he writes $n = k(a+b)$, such that

$$\Pr\left(\left|\frac{X}{n} - p\right| < \frac{1}{a+b}\right) > \frac{c}{c+1}$$

for some chosen positive constant c. The value of k, and hence n, is a function of a, b, and c. Nicolaus Bernoulli turned the problem around into finding bounds for the probability for a given value of n. Chapter I of Book V of *Miscellanea Analytica* is devoted to a review of the work of both Bernoullis. Motivated by their results or not, De Moivre definitely saw the connection. Following each of the first two problems, he gives corollaries in which he states that as more games are completed, the proportion of games won by the two players will be close to either $1/2$ each or to p and q as the case may be. De Moivre repeated these observations in more detail when he reproduced the first two problems in his second edition of *Doctrine of Chances*.

De Moivre showed at least some of his results in Book V, Chapter II to Cuming in the early 1720s and then sat on them for some time. Cuming seems to have been under no restriction to keep the results to himself. Around 1725, he told James Stirling about De Moivre's approximation to the middle term of $(1+1)^n$ divided by 2^n and challenged Stirling to find the result using finite difference methods.[21] Over time Stirling did respond to the challenge. He had the same problem as De Moivre in getting his research done and his results to print. Putting his research together and getting it published took time. He was planning his own book, *Methodus Differentialis*, as early as 1725, but it did not get it to print until 1730 due, like De Moivre, to his necessity to teach to make a living.[22]

Stirling wrote to De Moivre on June 19, 1729, with his solution to finding the middle term of the binomial.[23] It was in the form of an infinite series given by

$$\sqrt{\frac{2}{\pi}}\sqrt{\frac{1}{n+1} + \frac{1^2 A}{2(n+3)} + \frac{3^2 B}{4(n+5)} + \frac{5^2 C}{6(n+7)} + \frac{7^2 D}{8(n+9)} + \frac{9^2 E}{10(n+11)} + \cdots}$$

where

$$A = \frac{2}{\pi(n+1)}, \quad B = \frac{1^2 A}{2(n+2)}, \quad C = \frac{3^2 B}{4(n+4)}, \quad D = \frac{5^2 C}{6(n+6)},$$

and so on. For $n = 100$ and carrying the series to ten terms, Stirling obtained 0.0795892373872 for the probability of obtaining 50 successes in 100 trials.[24] This agrees with a modern exact calculation of the probability to twelve significant digits using R, a programming language. De Moivre's approximation in *Miscellanea Analytica* yields 0.0795559 to six significant digits.

De Moivre inserted Stirling's letter not after his work on the normal approximation to binomial probabilities in Book V, but instead near the end of his discussion of finite difference methods in Book VII, Chapter II where he replies to Montmort's accusations against him. In the preamble to the presentation of Stirling's letter, De Moivre discusses problems with convergence of series based on finite difference methods and how Stirling's methods have overcome this problem.

Within a few days of *Miscellanea Analytica* appearing early in 1730, and presumably Stirling receiving his copy as a subscriber, Stirling wrote again to De Moivre.[25] This time he informed De Moivre that the table of logarithms that De Moivre had calculated for use in his approximation to the binomial had errors in it. The table contains numerical values of $\log_{10}(x!)$ for values of x between 10 and 900 in steps of 10. Stirling also supplied De Moivre with a new result that he had obtained for an approximation to $x!$. As De Moivre explained it, the approximation is obtained from the series

$$z\log_{10}(z) - az - \frac{a}{2 \times 12z} + \frac{7a}{8 \times 360z^3} - \frac{a}{32 \times 1260z^5} + \frac{a}{128 \times 1680z^7} - \cdots$$

plus a constant, where $z = x + 1/2$ and $a = \ln(10)$ in the series. The constant is $(1/2)\log_{10}(2\pi)$ or $\log_{10}(\sqrt{2\pi})$. The series can be obtained from Proposition 28 in Stirling's *Methodus Differentialis*, which at the time was still in press. De Moivre would have received his copy late in 1730. The proposition deals with finding the sum of the logarithms of numbers which form an arithmetical progression. The earlier series sent to De Moivre in 1729 that approximates the middle term in the binomial is given in Proposition 23 of *Methodus Differentialis*.[26]

The numbers 12, 360, 1260, and 1680 in Stirling's new series were familiar to De Moivre. The number $2\frac{21}{125} = 2.168$, which was used in his approximation to the middle term in the binomial is obtained from De Moivre's derivation of an infinite series given by

$$\frac{1}{12} - \frac{1}{360} + \frac{1}{1260} - \frac{1}{1680} + \cdots.$$

In particular, $2e^{1/12 - 1/360 + 1/1260 - 1/1680} = 2.168208$ to six decimal places. How De Moivre obtained $2\frac{21}{125}$ is not given in Book V of *Miscellanea Analytica*. The details are saved for Book VI, which is devoted to a number of problems in infinite series. After seeing Stirling's result, De Moivre went back to his infinite series and reworked the problem. He obtained the series

$$\left(x-\frac{1}{2}\right)\ln(x)+\frac{1}{2}\ln(2\pi)-x+\frac{1}{12x}-\frac{1}{360x^3}+\frac{1}{1260x^5}-\frac{1}{1680x^7}+\cdots$$

for $\ln((x-1)!)$. His new results along with the corrected table were published in a supplement to *Miscellanea Analytica*. Presumably, it was printed at De Moivre's expense and sent to the subscribers who had picked up their books. In a corollary in the supplement, De Moivre notes that the series

$$1-\frac{1}{12}+\frac{1}{360}-\frac{1}{1260}+\frac{1}{1680}-\cdots$$

approximates $\ln\left(\sqrt{2\pi}\right)$ and attributes this insight to the series he had received from Stirling. In a corollary that immediately precedes this, De Moivre recognized that there were problems with the convergence of his series after the first five terms.

As a brief aside, Stirling's series from his Proposition 28 leads to

$$\sqrt{2\pi}\left(x+\frac{1}{2}\right)^{x+\frac{1}{2}}e^{-x-\frac{1}{2}}$$

as an approximation to $x!$. De Moivre's series leads to $\sqrt{2\pi x}\,x^x e^{-x}$. Today, the latter approximation is the one that is typically used and is almost always called Stirling's approximation.[27]

De Moivre gathered his conclusions together on the normal approximation to binomial probabilities, did some further work, and then put it all together in one document that he titled *Approximatio ad Summam Terminorum Binomii* $\overline{a+b}|^n$ *in Seriem expansi*.[28] It was published in November 1733. As in *Miscellanea Analytica*, he dealt initially with the case in which $a=b=1$. With Stirling's insight, he could now express M as

$$\frac{2}{\sqrt{nc}},$$

where c is the "Circumference of a Circle whose Radius is Unity," or in other words 2π. With one exception, De Moivre refers the reader to his *Miscellanea Analytica* for all the technical details requiring infinite series. The exception is a new approximation he obtained for finding the term that has $m+l$ successes, where

$m = n/2$. De Moivre showed that

$$\ln\left(\frac{Q}{M}\right) \cong -\frac{2l^2}{n}.$$

De Moivre went on to consider the general case in which $a \neq b$. There he found that the maximum in the expansion of $(a + b)^n$ occurs when "the Indices of the Powers of a and b have the same proportion to one another as the Quantities themselves a and b." This is the same as saying that maximum occurs when the power of a, or the number of successes, in the expansion of $(a + b)^n$ is $m = np$, provided that np is an integer. The value of M in the general case is

$$\frac{1}{\sqrt{2\pi npq}}$$

and the relationship between the maximal value and the value for $m + l$ successes, where $m = np$, is given by

$$\ln\left(\frac{Q}{M}\right) = -\frac{l^2}{2npq},$$

approximately.

De Moivre continued to interact with James Stirling as the decade of the 1730s progressed. In the late 1730s, Stirling was corresponding with Leonhard Euler about infinite series.[29] Stirling must have kept De Moivre informed about the correspondence. This is apparent in a letter written in July 1744, from De Moivre to Philip Stanhope.[30] When Stanhope and De Moivre last met in person, one of the topics of their discussion turned to Euler's method of evaluating the series

$$\frac{1}{1^2} + \frac{1}{2^2} + \frac{1}{3^2} + \frac{1}{4^2} + \cdots,$$

which sums to $\pi^2/6$. First considered in the mid-seventeenth century, it was not until 1736 that Euler became the first to obtain the result.[31] It is known as the Basel problem since Jacob Bernoulli tried to solve the problem; when he could not, he asked that anyone who found the solution should send it to him in Basel. During the face-to-face discussion with Stanhope, De Moivre told the earl that he had solved the Basel problem "some years ago." Stanhope replied that he thought the solution could be obtained through Newton's quadratures. Part of the letter is devoted to showing Stanhope how a combination of quadratures and infinite series methods from *Miscellanea Analytica* can be used to show that

$$\frac{1}{1^2}+\frac{1}{3^2}+\frac{1}{5^2}+\frac{1}{7^2}+\cdots$$

sums to $\pi^2/8$. If

$$\frac{1}{2^2}+\frac{1}{4^2}+\frac{1}{6^2}+\frac{1}{8^2}+\cdots=\frac{1}{4}\left(\frac{1}{1^2}+\frac{1}{2^2}+\frac{1}{3^2}+\frac{1}{4^2}+\cdots\right)$$

is added to De Moivre's series, then Euler's series is obtained. On denoting the sum of Euler's series by S we have the relationship $S = \pi^2/8 + S/4$, which yields $S = \pi^2/6$. De Moivre told Stanhope that he had heard of Euler's result from Stirling after Euler had written Stirling about it "about seven years ago." All that De Moivre learned from Stirling, besides the result, is that Euler had not given Stirling the complete proof to the result and that Euler "had done it by a Method peculiar to himself."

    Miscellanea Analytica and the *Approximatio* were written in Latin. From medieval times, Latin was the international language of diplomacy, law, scholarship, and the western part of the Christian church. It became the lingua franca of the Renaissance and remained the language of science into the eighteenth century. *Miscellanea Analytica* and the *Approximatio* were written with an international audience in mind. De Moivre's composition in Latin was done intentionally and began early in his career. For example, he used his 1704 *Animadversiones* criticizing Cheyne as an entry into correspondence with Johann Bernoulli and the wider Republic of Letters. As De Moivre stated in the second edition of *Doctrine of Chances*, De Moivre had the *Approximatio* printed for private distribution among his friends. In view of the language used, it was not only for his British friends (James Stirling's copy is held by University College London), but also for others on the Continent. There is a copy in the Basel University Library that is bound with *Miscellanea Analytica* and its supplement.[32] Since this book survives in its original binding, it may have been a copy obtained originally by some Swiss mathematician, finding its way eventually to the Basel library. Gabriel Cramer, as well as Johann and Nicolaus Bernoulli, are all possible candidates for the Swiss mathematician. All of them apparently received copies of *Miscellanea Analytica* and therefore would be likely recipients of the *Approximatio* as well.

    It is therefore of interest to divide De Moivre's publications into those written in Latin, and those in English. His work in natural philosophy, material on centripetal forces in particular, is published in Latin. This includes his 1717 paper on the subject in *Philosophical Transactions* and the fuller treatment in *Miscellanea Analytica*. All his research that deals with finding roots of equations using trigonometric methods based on divisions of the semicircle appears in Latin. The research is developed in papers published in 1707 and 1722 in *Philosophical Transactions*, as well as the material surrounding "De Moivre's theorem" in complex analysis in *Miscellanea Analytica*. Publications that are in English only deal with subjects of interest mainly to British mathematicians. These include, for example, generalizing Newton's binomial theorem to the multinomial situation; given a curve described by an equation involving a polynomial in *y* and a polynomial in *x*, solving for *y*; the study of a curve given in Newton's catalog of curves; and methods for the valuation of annuities.

De Moivre's work in probability appears both in Latin and in English. This deserves some further scrutiny. His first publication in probability is in Latin. In one sense it was an international showcase for his mathematical talents. Using the connections of his friend Pierre Des Maizeaux, De Moivre made sure that *De Mensura Sortis* was distributed at the highest level of the Académie royale des sciences in Paris. After 1711, De Moivre's probability publications are sometimes in English and at other times in Latin. By separating out the Latin publications, we can see that De Moivre was interested in having his international audience see what he thought was his most important work. From his discussions in 1717 with Brook Taylor on what he thought should be in the allegorical frontispiece in *Doctrine of Chances*,[33] De Moivre considered that he had made three important and highly original contributions to probability up to that point in time: the Poisson approximation to binomial probabilities; the solution to the problem of the pool for four or more players; and the trigonometric solution to the duration of play problem. The first appears in *De Mensura Sortis*. The second and third are in articles in Latin appearing in issues of *Philosophical Transactions* for 1714 and 1722, respectively. History would grant that his other major contributions are his development of generating functions (in Latin in *Miscellanea Analytica*), and his normal approximation to binomial probabilities and related material (in Latin in both *Miscellanea Analytica* and the *Approximatio*). All these results found their way into English in *Doctrine of Chances*. Finally, his response to Montmort's accusations of plagiarism was put in Latin in *Miscellanea Analytica* for any educated person, in Britain or on the Continent, to read and understand.

De Moivre continued to be the mathematics man for the Royal Society. In June 1733, the Royal Society received a gift of a book from Claude Richer du Bouchet.[34] Writing from Paris, Richer sent the book *Analyse générale ou méthodes nouvelle pour resoudre les problêmes de tous les genres* by Thomas Fantet de Lagny. Lagny had been a fellow of the Royal Society since 1718, sponsored by De Moivre. Lagny was in his seventies when the book was published. He may have been ill at the time since it was Richer who edited the material and brought the book to print. Richer's gift was accompanied by a request. According to the minutes of the meeting, Richer "expresse[d] a zealous inclination to be recommended to the Society in the capacity of a candidate for election." The Royal Society referred the book to De Moivre. Richer was never elected fellow of the Royal Society and, moreover, was never made a member of the Académie royale des sciences.

Following quickly on his mathematical triumphs of the early 1730s, De Moivre was struck by tragedy. It had started on a very promising note. His nephew, Daniel junior, was employed by Sir Joseph Eyles, a London merchant with powerful connections. Eyles was a director of the East India Company, the Bank of England and London Assurance. His brother, Sir John Eyles, was also a director of the East India Company and the Bank of England, as well as a director of the South Sea Company. When he was studying at Cambridge, Sir John's son Francis subscribed to *Miscellanea Analytica* and so was possibly a De Moivre student at one time. Daniel junior's job may have come through his uncle's connections. At some point very early in the 1730s, using South Sea Company connections, Daniel decided to break out on his own as an exporter of British luxury goods.[35] For his enterprise to work, Daniel needed the South Sea Company. Prior to 1713, Spain severely restricted trade between other countries, including Britain, and her colonies in the New World. By the Treaty of Utrecht that ended the War of Spanish Succession, the South Sea Company was allowed to send one ship a year to New Spain with a 500-ton cargo and was given the monopoly for supplying New Spain with slaves from Africa.[36] Eying this new market, Daniel convinced Sir Joseph Eyles and a few others, including his father Daniel senior, to underwrite an enterprise to export British-made jewelry and other luxury items to Veracruz in Mexico, at the time a part of New Spain.[37] As a port city and one of the two commercial centers of New Spain, Veracruz had a wealthy merchant class that could be interested in Daniel's merchandise. After obtaining a little more than £1400 in loans, including £700 from Sir Joseph Eyles and £90 from his father, Daniel junior went on a buying spree. He bought rings, earrings, watches, gold and silver chains, and silver boxes from several Huguenot and non-Huguenot artisans. He also kitted himself out for the trip with the purchase of several items of clothing, a cutlass, a field bed, and a hammock. In September 1732, Daniel junior set sail on the South Sea Company trading ship, the Royal Carolina for the British colony of Jamaica and then on to Veracruz. Prior to his departure he wrote his will, naming his wife, Marianne, as well as Sir Joseph Eyles and one other, as executors.

Then it all came apart. Daniel De Moivre senior died in 1733 at the age of about 64, followed by Daniel junior the next year at the age of about 27. Daniel junior had not sold all the goods he took to Veracruz. By order of the Prerogative Court of Canterbury, Sir Joseph Eyles became the sole executor of Daniel junior's estate.[38] The estate was not settled until 1738 or later. In May of that year, Daniel's creditors were informed that they were to come and substantiate their claims before one of the masters in Chancery in order to receive payment.[39] It is difficult to say what kind of financial straits the family was in. In his will, Daniel senior had bequeathed £200 to his son and the remainder of his estate, an unspecified amount, to his wife, Anne.[40] The newspaper announcement to Daniel junior's creditors makes no mention of bankruptcy. There may have been sufficient money to pay all the creditors, but it would have been tied up in the courts until they were all paid. Until Marianne De Moivre remarried in 1737,[41] it is likely that Uncle Abraham provided some financial support to Daniel junior's family. He may also have given some support to his sister-in-law Anne De Moivre until her death.

Daniel De Moivre junior was fairly close to his uncle. A list of his correspondents in 1732 and 1733 shows that he sent letters to his mother and father, his wife, his business associates, and his uncle Abraham. Despite the closeness, he had little interest in his uncle's work. Based on a list of the books that he owned at the time of his death, his reading habits ran to belles lettres like his uncle. But he did not own any books on mathematics. He also gambled in a moderate way, playing at cards. Keeping accounts in 1731 and 1732 in preparation for the voyage to Veracruz, he jotted down his wins and his losses for his card games. At any weekly game, his wins and losses were as much as £4. Typically, in a month he averaged about £1 on the win side. There is no mention in his papers of any gambling advice from his uncle.

The tragedy behind him, Abraham De Moivre was planning to publish the second edition of *Doctrine of Chances* as early as 1736. At that time he advertised for subscribers to the book. The price for a subscription was one guinea, or 21 shillings; and he planned to print only as many books as the number of subscriptions that he received.[42] This was similar to his plan for the publication of the first edition and for his *Miscellanea Analytica*.[43] This time he changed printers to one named Henry Woodfall, another well-respected printer.[44] At the time De Moivre was 69 years old. According to soon-to-be-released actuarial tables, he should have been very near the end of his life. Thomas Simpson's 1742 life table[45] shows only 75 survivors to age 69 out of 1280 births and only 29 survivors to age 80.

At the beginning of 1738, when *Doctrine of Chances* came off the press, De Moivre advertised in the *London Daily Post and General Advertiser* for subscribers to pick up their book from Slaughter's Coffeehouse. At least that is my conjecture. There are several issues of the newspaper from the beginning of 1738 that have not survived. An advertisement placed in the newspaper for July 11 and July 18 states that De Moivre had earlier advertised the availability of the book and that subscribers who had not yet picked up their copy could send for them at Slaughter's before the end of the year.[46] The two-year time span between collecting subscriptions and the final publication is similar to what occurred with *Miscellanea Analytica* and the known time between when he decided to publish the first edition of *Doctrine of Chances* and when it actually saw print.[47]

De Moivre took his accumulated results in probability since 1718, including material in *Miscellanea Analytica*, and revised what he had in the first edition of *Doctrine of Chances* by changing the wording and adding more examples and explanation to make the new edition of the book. He also included all his work on annuities. On that subject, De Moivre corrected on error that he had made in *Annuities upon Lives* when dealing with successive lives. This is an annuity contract that pays an annual amount for the life of the annuitant and then upon his or her demise for the life of a second-named person.

When the first and second editions of *Doctrine of Chances* are compared, the major additions of new results to the second edition from his published work are the generating function to obtain the probability of the sum of the faces that show on the throw of dice; the trigonometric solution to the duration of play problem; the expectation of the absolute deviation from the mean for the binomial distribution; and the normal approximation to binomial probabilities. It would be his magnum opus in probability, a complete compendium in English of the latest results in the theory and application of probability.

As the pièce de résistance to the new edition, De Moivre translated the *Approximatio* into English, intending it to appear at the end of the book. It did not quite work out that way. As the book was in press, he could not resist adding some new problems that had recently been posed to him. They were tagged onto the end of the book.

One of the new probability problems that De Moivre added as the book was going to press harks back to a challenge problem set by Nicolaus Bernoulli for De Moivre in 1714 or 1715. In the late 1720s, Bernoulli described the problem to Gabriel Cramer.[48] A and B play a game with a four-sided die with the sides marked 0, 1, 2, and 3. Both A and B put stakes, not necessarily of the same amount, into a pot. When they take their turns throwing the die each takes out of the pot a unit of money corresponding to the number shown on the die. There is an exception. When A throws the die and 0 shows, then he puts one unit of money into the pot; and if a 2 or 3 shows and the pot has less money than what shows on the die, then he puts into the pot the difference between what is in the pot and what shows on the die. The problem is to find what stakes A and B should each put into the pot so that it is a "fair game" in that each has the same expectation. Bernoulli intended this as a problem that could not be solved using the binomial theorem or infinite series. The problem suggested to De Moivre and solved by him in 1738 is easier—and it could be solved using infinite series.[49] A number of players each put equal stakes into a pot and then play with a four-sided die. The die is now marked T, P, D, and A rather than 0, 1, 2, and 3. The players throw the die in turn until the pot is won. If T shows on a throw, the player throwing it wins the pot. If P shows, then the player adds to the pot so that it doubles in size. If D shows, there is no change to the size of the pot. Finally, if A shows, then the player takes half the pot. At the completion of a throw, the turn passes to the next player. The problem is to find the expected winnings of each player.

The solution to another problem made further inroads into the use of generating functions to solve probability problems. It was De Moivre's solution to the problem of runs.[50] A run of successes of length r in a sequence of n trials, each of which may result in success or failure, is any set of r successes in a row. It is usually assumed that the trials are independent and that there is a constant probability of success p from trial to trial. De Moivre states the problem and gives the solution through a generating function. The expansion of the generating function can be done using one of De Moivre's results in recurring series. His discussion of the problem contains some numerical examples but no further proof or explanation of how he obtained the generating function.

Over a century later, Isaac Todhunter explained how the generating function could be obtained through a recurrence relationship.[51] First, a run of length r in a series of n trials can be obtained by a run of r successes in $n-1$ trials irrespective of what happens on the nth trial, or by not obtaining the run in the first $n-r-1$ trials followed by a failure and then r successes in a row. Consequently, the run probability for r successes in n trials is the sum of the run probabilities involving r successes and $n-1$ trials and r successes and $n-r-1$ trials.

Not having access to Todhunter's book, which was 120 years into the future, De Moivre's former student Philip Stanhope, 2nd Earl of Stanhope, read the second edition and attacked the problem of runs in his own way in the 1740s. He recognized that a run of exactly r successes in n trials can occur in those n trials by having r successes and $n-r$ failures, or $r+1$ successes and $n-r-1$ failures, or $r+2$ successes and $n-r-2$ failures and so on. The required run probability, which Stanhope successfully obtained, is then a weighted sum of the probabilities of these events of successes and failures. In correspondence with Stanhope, Thomas Bayes also attacked the problem of runs, but his solution is incorrect.[52]

From a modern viewpoint, the 1736 advertisement for the proposed new edition of *Doctrine of Chances* overstates the claims about what De Moivre had done in his book. It is likely that Pierre Des Maizeaux wrote the advertisement. The advertisement was part of a report on new publications coming out of London and Des Maizeaux was one of the active writers for *Bibliothèque raisonnée*, where the report appeared. In the advertisement, the question is asked, Can we reasonably conjecture based on past experience? The writer goes on to answer his own question in the affirmative, but was incorrect. His answer was that while single events might be unpredictable, there is stability and order in the long run. Writing in 1749, the philosopher and physician David Hartley made the distinction very clearly between two different questions and the applicability of their answers.[53] In his *Observations on Man*, Hartley saw De Moivre's result on the stability of long-run frequencies "as an elegant method of accounting for that order and proportion, which we every where see in the phænomena of nature." The question in the advertisement was another one altogether, which Hartley calls "the inverse problem." Given that a ratio of successes to failures has been observed in the past, what is the probability that the observed ratio deviates from the true ratio by any given amount? This is the question that Thomas Bayes answered satisfactorily in his posthumous publication in 1763.[54] Hartley attributed the solution only to "an ingenious Friend," which prompted Stephen Stigler to assess the evidence about who this friend might be.[55] Stigler came to two possibilities: Nicholas Saunderson and Thomas Bayes. New evidence that might tip the balance in favor of Bayes is that most of Bayes's known mathematical manuscripts date from the 1740s.[56]

✥ 13 ✥
The Two Thomases

This is a tale of two Thomases: Thomas Simpson and Thomas Bayes. Each appears on the mathematical scene, seemingly independently, late in De Moivre's career. They were both influenced by De Moivre's work, especially his work in probability and its applications. For his part, De Moivre was unhappy with some of the work of each Thomas, more so that of Thomas Simpson. These are also stories of a connection broken and a connection never made. After a cordial beginning, De Moivre had a distinct falling out with Thomas Simpson. Although he probably knew Bayes, or knew of him, Bayes was apparently never part of De Moivre's closer circle of friends and colleagues.

There are some similarities between the two mathematicians named Thomas. And they probably knew one another by the late 1740s. Both had patrons, or at least very helpful individuals, who were formerly De Moivre's students. Simpson obtained a position teaching mathematics at the Royal Military Academy at Woolwich in 1743 on the recommendation of Martin Folkes. It was a new position; the Academy was founded only two years earlier in 1741. Bayes probably met Philip Stanhope, 2nd Earl Stanhope, on one of Stanhope's visits to Tunbridge Wells, where Bayes was the minister at the dissenting chapel. It was a popular spa town for the wealthy. Subsequently, they corresponded and when Bayes was put up for fellowship in the Royal Society, Stanhope was his first sponsor. Both Simpson and Bayes were elected fellows of the Royal Society in the first half of the 1740s, Bayes in 1742 and Simpson three years later in 1745.[1] They were elected on the basis of their mathematical interests. Both had a common friend in another Royal Society fellow, John Canton; through Canton, Bayes commented on a paper by Simpson.[2] No confirmed portrait of either Bayes or Simpson is known to survive and both died in 1761, within two months of each other.[3]

There are also some significant differences between the two Thomases. Bayes was moderately wealthy, while Simpson initially struggled before he obtained his patronage position. Simpson was prolific in his mathematical writing; Bayes wrote very little and had his major work published posthumously. Simpson had little formal education and was self-taught in mathematics. Bayes may have been tutored in his youth or entered a dissenting academy in London, and then entered the University of Edinburgh, where he prepared for the Presbyterian ministry. Along the way at Edinburgh, he studied mathematics under a lesser known member of the Gregory family of mathematicians.

Thomas Simpson

When De Moivre and Simpson first met, it was at a dinner. Their initial meeting was cordial. De Moivre was his usual "très joyeuse compagnie" as Jean Des Champs described him in the late 1740s.[4] When introduced to Simpson, De Moivre held open his arms and is reported to have said, "I am delighted to see you, I honour your talents and embrace you with all the vivacity of a Frenchman and the sincerity of an Englishman."[5] Before dinner, they talked mathematics. De Moivre gave Simpson a problem to work on at his leisure. As described by the French astronomer Joseph Jérôme de Lalande, it required integrating a complicated differential equation. Put in its historical context, this is the continental language of calculus. De Moivre probably challenged Simpson to find the quadrature of a curve defined by a fluxional equation or to find a certain fluent from its fluxion. The meeting could have occurred in the mid-1730s as Simpson was gaining a reputation as a tutor in mathematics.

By the early 1740s their relationship had soured considerably. Their estrangement is evident in the exchanges between them in De Moivre's 1743 edition of his *Annuities on Lives* and Simpson's response to it that same year. Although the dispute has been described by historians of statistics such as Stephen Stigler and Anders Hald through De Moivre's and Simpson's publications on annuities,[6] a slightly fuller story emerges when other source material is considered.

The conflict between De Moivre and Simpson has its origins in the publication of De Moivre's second edition of *Doctrine of Chances*, De Moivre's magnum opus in probability. A work of this caliber, arguably by Britain's then preeminent mathematician, attracts some of what might be called riding-on-the-coat-tails activity. John Arbuthnot's *Laws of Chance* was resurrected, but not Arbuthnot himself to do a new edition. The printer for Arbuthnot's original edition of 1692, and a subsequent one in 1714, was Benjamin Motte, a prominent printer and bookseller who published several works of Arbuthnot's friends in the Scriblerus Club, including *Gulliver's Travels* by Jonathan Swift. Arbuthnot died in 1735, so Motte gave responsibility for the revision of *Laws of Chance* to John Ham, a mathematics teacher at a school in Hatton Garden located about half a mile from Motte's shop. By this time Motte was in partnership with Charles Bathurst. A third, probably pirated, edition came out in

1731, of which no copies seem to have survived. It was printed by James Roberts.[7] The third edition may have been a coat-tail effect of the publication of *Miscellanea Analytica* in 1730. Although Motte and Bathurst's fourth edition of *Laws of Chance* bears the publication date of 1738, advertisements for its availability at their shop appear in mid-September, 1737, just prior to the release of *Doctrine of Chances*.[8] These advertisements continued until March 1738.[9] The fourth edition contains all of Arbuthnot's original material, primarily an English translation of Huygens's *De ratiociniis in ludo aleae*, and Arbuthnot's analysis of some games of chance. Ham added new material, some taken directly from De Moivre's first edition of *Doctrine of Chances* and some containing analyses of games not considered by De Moivre. Ham also considered De Moivre's problem of finding the number of trials required to obtain, with probability $1/2$, at least two successes in a series of independent trials whose outcomes are success or failure. While De Moivre assumed that the number of trials was large and the probability of success small to get what is known as the Poisson approximation to the binomial, Ham was able to get an infinite series approximation to the solution that relaxed this assumption.[10] Ham's update of Arbuthnot's book was no real threat to De Moivre's work or to his new book, which had many new, additional important results.

On July 11 and July 18, De Moivre placed an advertisement in a London newspaper informing the subscribers to *Doctrine of Chances* that those who had not yet picked up their copy could send for them at Slaughter's Coffeehouse by the end of the year.[11] Then three days after this advertisement last appeared, on July 21 a very odd advertisement appeared in the same newspaper:

Mr. SIMPSON Defended.

THO' it has been ungenerously replied to a late Advertisement, in favour of Mr. *Simpson*, and His New Treatise on the Doctrine of Chances, that such was scandalous, and evidently intended to injure him; it is thought proper to advertise, that no Harm was meant him at all, unless a good Opinion of his Abilities was Harm. What is meant by *Bare-fac'd Piece of Villainy* is not understood, unless the purchasing so valuable a PIECE, at so undervalued a Rate. The Author of the abus'd Advertisement professes Friendship to Mr. Simpson, and therefore could intend nothing but what is Honest. He will make good every Assertion of his *New Discoveries*, notwithstanding what has been enviously denied by a Person who meant to deny him the Honour. He agrees that the *summing up of Series* is something Curious, since a Method from thence may be deduc'd for finding the *Superficial Content* of any Writer, (a Curve Superficies never before attempted) and had the Writer of the Mock Advertisement (on Monday last) been sensible of the Praise he bestow'd, instead of what he intended, he would have been silent in several Particulars.

Button-Court, near the Monument.   N. READ[12]

I have scoured the surviving London newspapers printed a month or more before this advertisement appeared and could find nothing to which the advertisement might refer. There are some clues that this advertisement may be a plant mocking Simpson's intended book. For example, I have been unable to locate a Button Court in eighteenth-century London, even on a 1756 London map of the area around the monument to the Great Fire of London.[13] The writer identified by "N. Read" might be a statement not to read Simpson's book rather than to a real person. Finally, the newspaper was operated and printed by Henry Woodfall, De Moivre's printer for the second edition, so Woodfall may have been trying to protect his author by allowing the advertisement to run. The reference to "purchasing so valuable a PIECE, at so undervalued a Rate" refers to Simpson's about-to-appear *Nature and Laws of Chance*. When it was advertised for sale in January of 1740, the book was priced at three shillings.[14] This is a substantial difference from twenty-one shillings—the price of De Moivre's book.

Beyond the price, what may have upset De Moivre even more is the content of Simpson's *Nature and Laws of Chance*. Unlike Ham's update to the *Laws of Chance*, Simpson's book is generally a cheap knock-off of the probability part of De Moivre's second edition of *Doctrine of Chances*. Simpson went through most of De Moivre's problems and provided one or two new ones of his own. For the most part, Simpson's solutions parallel or repeat De Moivre's and are sometimes simpler.

There are a couple of exceptions where Simpson provided some of his own unique solutions. In the problem of finding the number of trials required to obtain, with probability $1/2$, at least two or more successes in a series of independent trials, De Moivre appears to use a simple numerical technique called the method of false position to obtain a numerical solution. Earlier in *De Mensura Sortis*, De Moivre stated that the results could be obtained through a power series in the fluent y evaluated at $y = \ln(2)$. Simpson obtained a general infinite series solution to the problem, not specifically in $\ln(2)$, when the number of trials is large and the probability of success is small.[15] Simpson also attacked the problem of runs in a way that is different from De Moivre's approach. In *Doctrine of Chances* De Moivre stated the problem and gave the solution through a generating function. His discussion of the problem contains some numerical examples but no further proof or explanation of the generating function. Simpson provided an explicit series solution to the problem.[16]

Simpson's publisher was Edward Cave. He seems to be the equivalent today of a mass-media publisher. He published *Gentleman's Magazine*, whose circulation reached about 15,000 subscribers.[17] At the price of three shillings, Simpson's book was meant for the general public, the "mass audience" that Cave pursued. Since De Moivre's book was sold by subscription and he controlled the distribution of it, De Moivre intended his book for a select audience. Even Simpson recognized this in his preface to *Nature and Laws of Chance*.

Despite the price differential and a mock advertisement that could not be attributed directly to De Moivre, or to any of his friends, no ill will seems to have broken out into the open at this time.

In 1742, Simpson published his *Doctrine of Annuities and Reversions*. For this book he had a different publisher—John Nourse. Where De Moivre relied on his model of linear survivorship based on Halley's table to carry out annuity valuations, Simpson was more closely tied to the actual life table. He tried to obtain a life table more applicable to London since Halley's table is based on data from a small city in what is now the southwest of Poland.

Simpson was very adept at annuity calculations and always seems to have had his eye on the practical. He checked De Moivre's approximation to joint-life annuities that relies on the incompatible combination of linear and exponential survivorship. To do this, he calculated the annuity values from his life tables, values based on the linear assumption alone, and values based on De Moivre's assumption. For the single-life annuity values that he considered, Simpson found that for two lives, De Moivre's approximation undervalued joint-life annuities by between 0.7 and 0.9 years purchase and that the linear assumption overvalued the same annuities by between 0.1 and 0.3 years purchase. Clearly to Simpson, the simple linear assumption was a better approximation. This contradicted De Moivre's very brief analysis of his approximation in the 1725 edition of *Annuities upon Lives*.

Simpson, however, may not have been that careful with his own annuity calculations. I have compared his table that shows a_x, the value of a single-life annuity issued at age x, to the value calculated using his life table.[18] The graph shows the differences between these values for ages at issue 6 through 75 at rates of 3%, 4%, and 5% interest. The differences are shown in years purchase and normally differ by 0.2 or 0.3 years purchase with some larger differences at the higher and lower ages.

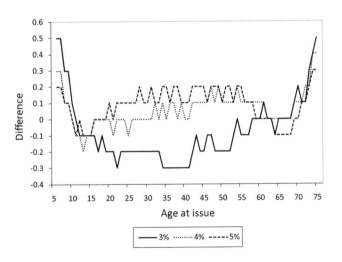

Differences between Simpson's tabular values of a_x and the values computed from his life table.

De Moivre responded with a second edition of his annuity book in 1743. The title was changed slightly to *Annuities on Lives*. He ignored Simpson's study of the accuracy of his approximation to the value of joint-life annuities. He preferred the simplicity of his approximation to some of Simpson's approaches. Concerning Simpson's whole approach to annuities, De Moivre wrote in his preface,

> After the pains I have taken to perfect this Second Edition, it may happen, that a certain Person, whom I need not name, out of *Compassion to the Public*, will publish a Second Edition of his Book on the same Subject, which he will afford at a *very moderate Price*, not regarding whether he mutilates my Propositions, obscures what is clear, makes a Shew of new Rules, and works by mine; in short, confounds, in his usual way, every thing with a croud of useless Symbols; if this be the Case, I must forgive the indigent Author, and his disappointed Bookseller.[19]

Whether or not his approximation was accurate, De Moivre's point about "obscures what is clear" and "makes a Shew of new Rules" is illustrated in their two approaches to the valuation of joint-life annuities. De Moivre's valuation of a joint-life annuity for two lives aged x and y is given by

$$a_{xy} = \frac{(1+i)a_x a_y}{\left(1+a_x\right)\left(1+a_y\right)-(1+i)a_x a_y},$$

where i is the rate of interest. At the end of his book he provides tables for a_x at various ages of issue, as well as tables for the value of fixed-term annuities, so that the final calculation is simple and straightforward. Here is Simpson's procedure for the same problem:

Case I.

If the two lives be equal; enter tab. II. with the common age, and against it you will have the value required.

Case II.

If the given ages be unequal, but neither of them less than 25, nor greater than 50 years; take half the sum of the two for a mean age, and proceed as in Case I.

Case III.

If one or both ages be without the limits abovemention'd, but so that the difference of the values corresponding to those ages, be not more than 1/3 of the lesser; let 4/10 of that difference be added to the lesser value, and the sum will be the value sought.

Generally,

> Be the difference of the values what it will, multiply it by 1/2 the lesser of the two values, dividing the product by the greater; then the quotient, added to the lesser value, will give the true answer very near.[20]

There are similar complicated rules for the valuation of last-survivor annuities on two lives, as well as joint and last-survivor annuities on three lives.

De Moivre's new edition of *Annuities on Lives* would have been simple and fairly quick to put together. He lifted the material on annuities from the second edition of *Doctrine of Chances* and, with some changes in notation, put it into his new book. This included his derivations based on the incompatible assumptions of exponential and linear survivorship—he liked the simplicity of it and believed the approximation to be accurate. Then he added some new material. One addition is a revision to the tables accompanying the book. In the second edition of *Doctrine of Chances*, De Moivre had provided tables for single-life annuities and annuities for a fixed term at a 5% rate of interest. Here he added tables at a 6% rate of interest to his *Annuities on Lives*. The second addition is a result on how to value annuities when the payments are semi-annual.

Simpson responded with a pamphlet entitled, *An Appendix, Containing Some Remarks on a Late Book on the Same Subject, with Answers to some Personal and Malignant Misrepresentations, in the Preface thereof.* Published near the end of May 1743,[21] it was printed and sold by Nourse for a sixpence. Simpson replied to all of De Moivre's criticisms of his work and went on to point out some errors in the first and second editions of De Moivre's book on annuities. He concluded with: "Lastly, I appeal to all mankind, whether in his treatment of me, he has not discover'd an air of self-sufficiency, ill-nature, and inveteracy, unbecoming a gentleman."[22] Rather than turning the other cheek, Simpson slapped right back at De Moivre.

De Moivre's friend William Jones took exception to what Simpson had written. In the mid-1770s, Reuben Burrow was dining with John Robertson. Burrow was a mathematician and one-time assistant to Astronomer Royal Nevil Maskelyne; Robertson, a fellow of the Royal Society since 1741, was at the time clerk and librarian to the Royal Society. They dined at the Royal Society. Over dinner, as Burrow wrote in his diary, Robertson recollected an incident from over thirty years before. He told Burrow that Jones treated "his mathematical friends with a great deal of roughness and freedom." He went on to say that "he rated Mr. Thomas Simpson in such a manner about his paper against De Moivre, that Simpson said he would never go see to see him more, but he did again however."[23]

Burrow did not state when this occurred, but it was probably soon after the *Appendix* was published. This would have been about a year and a half before Jones became Simpson's first sponsor for his election to fellowship in the Royal Society.

Some may think that this is a fight simply about competition from a cheap knock-off. The problem with this interpretation by itself is that until 1743, De

Moivre had not lost any money on his books. *Doctrine of Chances* was sold by subscription and the number of copies printed was covered by the subscription. The price of three shillings for *Nature and Laws of Chance* may not have been that relevant to De Moivre in terms of competitive pricing. It was the publication of Simpson's annuities book that brought the conflict into the open. It is useful to separate *Doctrine of Chances*, which includes work on annuities, from *Annuities on Lives*. This will divide De Moivre's dispute with Simpson into economic and non-economic parts.

For the economic part of the argument, I begin by looking at the publishing history of some of De Moivre's books. So far, both editions of *Doctrine of Chances* were published by subscription. The 1725 *Annuities upon Lives* was not. It was printed by William Pearson and sold in the bookshops of Francis Fayram, Benjamin Motte, and William Pearson. Three years after its publication, Fayram was still advertising copies for sale.[24] Though influential, it was not an immediate runaway bestseller. The cost was three shillings. This was the same price as Simpson's 1742 *Doctrine of Annuities and Reversions*.[25] De Moivre's 1743 *Annuities on Lives* was printed for the author by Henry and George Woodfall. Unlike Simpson's books, I have found no newspaper advertisements for its sale by a bookseller or for its availability from the author. De Moivre must have sold the book himself to his network of friends and colleagues or to his clients who came to him for consultations regarding the pricing of annuities.

De Moivre came out with a third edition of *Annuities on Lives* in 1750 with the offending paragraph against Simpson removed. It was printed and sold by Andrew Millar, a leading London printer who was known to deal generously with his authors.[26] When Millar first advertised the sale of the third edition, the advertisement said "Price 2s. 6d. formerly sold for 5s." In the same advertisement, he offered for sale *Miscellanea Analytica* and the second edition of *Doctrine of Chances*.[27] Since both were published by subscription, these must have been extra copies that De Moivre still possessed and passed on to Millar. *Miscellanea Analytica* sold at ten shillings sixpence, half the original subscription price, and *Doctrine of Chances* sold at the full price of one guinea.

I have carefully compared the second and third editions of *Annuities on Lives* that are in electronic format on Eighteenth Century Collections Online (ECCO). Stephen Stigler also did me the favor of examining original copies in his possession. With the exception of the title page, epistle dedicatory, preface, and pages 59 to 68, the type looks to be identical between the two editions. This includes all the letters and numbers, the placing of these letters and numbers on the page, and the size and lengths of mathematical characters such as equal signs and minus signs. Further, the errata in the second edition have not been corrected for the third. Stigler also noticed that the paper used for the changed pages is lighter than the pages where the print is identical between the editions. Millar produced a fourth edition of *Annuities on Lives*. It was advertised for sale late in 1752 and was priced at two shillings sixpence.[28] The font in the fourth edition is the same as that used in the second and third editions, but the type throughout the whole book has definitely been reset. The content is the same as the second and third editions with the addition of annuity tables at 3 and 3.5%.

I am left with the distinct impression that De Moivre was unable to sell many of his copies of the second edition of *Annuities on Lives*, which was probably priced at five shillings, two more than Simpson's book. Here is a plausible reconstruction. Now in his eighties, De Moivre took all his leftover books to his new printer. His copies would not have been bound; bookbinding was separate from printing at this time. Millar printed a few new pages for *Annuities on Lives*, perhaps two full sheets meant for octavo pages or thirty-two pages in total, and replaced some of the pages in the remaining copies of the second edition to make the third edition. *Miscellanea Analytica* and the second edition of *Doctrine of Chances* were sold as is. Millar quickly sold out his copies of the third edition of *Annuities on Lives* and was then able to print a fourth to meet the demand for the book at the new price.

There could be other explanations for the transition from the second to the third edition of *Annuities on Lives*. One possible explanation is that De Moivre owned the type for the second edition and kept it intact. There would have been about sixteen trays of type. Since it was a mathematical book, Woodfall may not have had any further use for the type once he finished printing the book. The main argument against this is that type was expensive to manufacture, so it was usual to break up the type once a single sheet was printed. At most, De Moivre could have obtained the type for some of the mathematical expressions that were of no future use to his printer.

With the publishing background in place, we can now look directly at the economic part of the argument between De Moivre and Simpson. Before Simpson came along with his book on annuities, there were only a few books on the market that provided tables and methods to price annuities, usually related to land ownership. They relied on De Moivre's methodology, and often stated it in print, or on other questionable methods. Simpson provided a new and valid approach to annuity valuations, backing it up with mathematical arguments. This gave him credibility in the marketplace for any consulting activities that he might carry out for annuity pricing.

Many mathematicians used the writing of commercial arithmetic books to advertise or legitimize their profession as teachers or consultants. As teachers, they also used their own books in their classrooms. A good example of this kind of activity prior to the 1730s and 1740s is Richard Hayes. He authored several commercial mathematics books, including sets of interest tables and a book called *An New Method of Evaluating Annuities on Lives*. After teaching in someone else's employ or in partnership with another for several years, in the early 1720s he opened his own school that trained students in commercial mathematics.[29] His books were often printed for the author, and newspaper advertisements for his book often mention his school. On the title pages of his books he describes himself as, "Accomptant and Writing Master."

Put in this context, we can see the second and subsequent editions of De Moivre's *Annuities on Lives* in an economic light. His business as an annuity consultant was expanding and the printing of *Annuities on Lives* gave his consultancy enormous credibility. The new edition of *Annuities on Lives* came out because he needed to keep pace with Simpson, who seems to have begun providing advice on annuity-related material in the early 1740s.[30]

I think there is also a non-economic side to De Moivre's initial annoyance with Simpson that eventually exploded into print. De Moivre had shown this side of his character in his disputes with Cheyne and Montmort. During the Cheyne dispute, John Flamsteed commented how upset De Moivre was prior to the publication of Cheyne's book on inverse fluxions. The book was on a subject in which De Moivre had recently published. Since his annoyance began prior to publication, De Moivre must have seen Cheyne's work in manuscript form. He seems to have been concerned about the correctness of Cheyne's results and the quality of his mathematics. In his dispute with Montmort, De Moivre seems to have been stung by Montmort's accusations of plagiarism and claims that all of his work was derivative from what Montmort had done. His response to Montmort was to keep any new and important results that he obtained away from Montmort as much as possible. De Moivre's response to Simpson might be rooted in these previous intellectual altercations. De Moivre knew that Simpson's work was entirely derivative of more than twenty years of his own labor. And his work was sometimes copied in a way that he considered clumsy and inelegant.

Nothing more came of the argument between De Moivre and Simpson. As Lalande recounts, however, De Moivre was going to respond again to Simpson's preface in *Doctrine of Annuities and Reversions*, but was convinced by his friends not to do so.[31]

Thomas Bayes

There has been some thought that Thomas Bayes studied the mathematics of probability with Abraham De Moivre. This is a thought that grew in John Holland's mind, which he put on paper for his own lengthy biography of Bayes in the early 1960s[32] after reading George Barnard's earlier brief biographical notes on Bayes.[33] Barnard had speculated that Bayes learned his mathematics from De Moivre. We now know that Bayes studied mathematics at the University of Edinburgh.[34] The thought that Bayes learned probability from De Moivre may still continue. Banish that thought. I will argue that Bayes's entry to probability was through Philip Stanhope and it probably resulted in a little friction at a distance between Bayes and De Moivre. Rather than Bayes studying with De Moivre, my take on the situation is that they had little to do with one another—a connection never made. I will concede that Bayes was influenced by De Moivre through reading the second edition of *Doctrine of Chances* published in 1738.

It is Philip Stanhope who was responsible for Bayes's election to fellowship in the Royal Society. It is also probably Stanhope who encouraged Bayes to work in probability. Stanhope's family estate at Chevening was only about 15 miles from Tunbridge Wells. He definitely visited the spa town in 1736.[35] The two may have met on, or shortly after, this visit. In 1736, Stanhope would have been 22 years of age and Bayes about 35. They had at least one other common interest besides mathematics. Stanhope was a pious person who was interested in theological questions.[36] Bayes and Stanhope become relatively close. In 1747, Stanhope paid a two-guinea subscription to Bayes's Presbyterian chapel in Tunbridge Wells.[37] An aristocrat supporting a dissenting meetinghouse was not unheard of, but it was unusual.

In the late 1720s or early 1730s, entry into the Royal Society required recommendations from three fellows. There was also the requirement that the prospective fellow "send in specimens to show what part of philosophy [he is] particularly conversant."[38] If this policy continued into the 1740s, then Bayes was required to submit some mathematical paper, perhaps not an original work but something that displayed his mathematical knowledge. A good candidate for this paper is one that Bayes wrote and sent to Stanhope. A copy of it is in Stanhope's hand.[39] The object of the paper is to find the quadratic factors of the expression

$$1 - 2\cos(\theta)x^n + x^{2n},$$

where θ is some angle between 0 and 2π. Since the quadratic factors are polynomials of degree two, the factors can contain three terms. Consequently, the problem is often expressed, as Bayes did, of finding the "trinomial divisors" of the original expression. This was a problem that De Moivre had considered and solved in his 1730 *Miscellanea Analytica*. De Moivre's solution relies on geometric arguments and begins with the equivalent of what is now known as De Moivre's theorem or formula. In his 1742 *A Treatise of Fluxions*, Colin Maclaurin provides a different geometrical solution and follows it with another one that uses only algebra.[40] While still geometrical in nature, Bayes had yet another approach to solving the problem. I once argued that there was no apparent purpose to Bayes obtaining yet another solution unless it predates Maclaurin's.[41] A 1741 or early 1742 dating of Bayes's manuscript therefore seems appropriate and corresponds to Bayes's entry into the Royal Society.

Whether I am right or wrong that Bayes's paper on trinomial divisors provides an illustration to the Royal Society of his competency in mathematics, the work seems to have piqued Stanhope's interest in the problem and De Moivre's solution to it. In 1744, Stanhope was in conversation with De Moivre about the problem, questioning De Moivre's solution. On July 5, 1744, De Moivre wrote to Stanhope reiterating that the appropriate factors of the expression are of the form $1 - 2bx + x^2$

where b is the cosine of an angle.[42] Using a simple algebraic argument, De Moivre showed that this quadratic expression could not be factored further. He did this by showing that the equation $1 - 2bx + x^2 = 0$ has imaginary roots given by $b \pm \sqrt{b^2 - 1}$. Since b is the cosine of an angle, $b^2 - 1$ is the negative of the square of the sine of the same angle. De Moivre wrote again to Stanhope a week later with a detailed algebraic example, demonstrating the case in the original expression when $n = 3$. You can almost feel De Moive's frustration, yet careful patience, with his aristocratic former pupil. After he set out his derivations for Stanhope, De Moivre wrote, "but I believe your Lordship will be of opinion that this method of process is infinitely inferiour to the method I took before to prove the thing universally."

His methods of proof in *Miscellanea Analytica* still seem to have been questioned by Stanhope. An undated letter from De Moivre that follows the first two opens:

> After due examination of my fifth Corollary, I freely confess that the Transition from my Lemma to that Corollary was a little too sudden which occasioned some obscurity in it, and that althõ I have endeavoured in my former Letters to your Lordship to remove the Doubts that may be entertained about it, yet I am not entirely satisfied with what I have done, therefore permit me once more Mylord to try whether I shall this time be more successful, I hope your Lordship will forgive my dwelling so long upon the same subject, which fault proceeds from the earnest desire I have to clear up my thoughts to your Lordship to whose Judgment I shall be proud to submit whereof I have written.

Once again, De Moivre went on to explain his proof through some simple examples. This final letter may have been prompted by a possible meeting between Stanhope and De Moivre on July 24, 1744. On that day, Stanhope recorded in his expense account book that he spent one shilling at Slaughter's Coffeehouse, his only recorded visit to the place.[43]

Re-enter Bayes. At some point, probably in the mid-1740s, Bayes and Stanhope were reading through De Moivre's second edition of *Doctrine of Chances* and corresponding about their own solutions to problems in the book. For example, Bayes sent Stanhope his solution to the problem of runs, which De Moivre had considered in his book.[44] Once again, De Moivre had only stated the solution through a generating function without showing how he obtained it. And there is a minor typographical error in the solution. Bayes gave Stanhope his own solution to the problem. And it is incorrect.

What may have annoyed De Moivre is Bayes's other response to material in *Doctrine of Chances*, specifically the normal approximation to the binomial. This was De Moivre's approximation to the ratio of the middle term in the expansion of $(1 + 1)^n$ to 2^n, which involves the series $1/12 - 1/360 + 1/1260 - 1/1680$ to four terms. Using these numbers provides a good numerical approximation to the ratio.

In 1747, Bayes sent Stanhope a manuscript that opens with

> It has been asserted by several eminent Mathematicians that the sum of the Logarithms of the numbers 1. 2. 3. 4. 5 &c to z is equal to $(1/2)\text{Log},c + \overline{z + (1/2)} \times \text{Log},z$ lessened by the series $z - 1/(12z) + 1/(360z^3) - 1/(1260z^5) + 1/(1680z^7) - 1/(1188z^9) + \&c$ if c denote the circumference of a circle whose radius is unity.[45]

At some point, a copy of this manuscript was sent to John Canton, with some sentences added to the beginning and end of it. Canton eventually had it published in 1763, two years after Bayes's death.[46] Regarding the series that Bayes gave with $z = 1$, De Moivre had stated in *Doctrine of Chances*, "I perceiv'd that it converged but slowly." Bayes spent the rest of his manuscript showing that the series in z diverges. In another manuscript that he sent to Stanhope, Bayes found his own approximation to the factorial using a different infinite series argument and applied it to approximating the ratio of the middle term in the expansion of $(1 + 1)^n$ to 2^n. Ultimately, it was the same approximation that De Moivre had obtained. If Stanhope had informed De Moivre of Bayes's work, no doubt it would have annoyed De Moivre in the same way Stanhope had done by questioning the work on trinomial divisors in *Miscellanea Analytica*. De Moivre's annoyance with Bayes may have gone further. Bayes's demonstration that De Moivre's infinite series diverged was also an indication of a flaw, perhaps a minor one but in a major topic, in De Moivre's magnum opus. The normal approximation to the binomial was one of De Moivre's greatest achievements, and it was perhaps slightly tarnished.

There is no other surviving historical evidence thus far that connects Bayes and De Moivre. Rather, the evidence that does survive closely connects Bayes and Stanhope; and there is a hint in other surviving evidence that Bayes and De Moivre never met. Prior to his election, Bayes was first brought to a Royal Society meeting on March 25, 1742. Only fellows and their guests could attend a meeting and the attendance of non-fellows was recorded in the minutes. Bayes was elected fellow on November 24, 1742.[47] By this time De Moivre's attendance at meetings appears to have been minimal. On February 10, 1743, Roger Paman was nominated to the fellowship. De Moivre, Robert Barker, and George Lewis Scott signed the nomination form as his sponsors. Of the three, it is likely only Scott attended the meeting. It was he who presented to the Society a book by Paman on fluxions, likely to demonstrate Paman's suitability for election.[48] The next year, on June 7, 1744, it was William Jones who presented some of De Moivre's new work on annuities. The results were in the form of a letter to Jones that he brought to the meeting. There is nothing in the minutes to suggest that De Moivre was present. From all the surviving evidence, we can conclude only that Bayes knew of De Moivre's work, and perhaps vice versa. We can also reasonably speculate that Bayes and De Moivre had little, if any, personal interaction.

∽ 14 ∾
Old Age

At the age of 72, Abraham De Moivre finally got his chance at an academic position in England. Following the death of Nicholas Saunderson on April 19, 1739, the position of Lucasian Professor of Mathematics at Cambridge became vacant. Prior to becoming Master of the Mint, Newton had held the professorship from 1669 to 1702. In order to be eligible for the position, De Moivre was given by royal warrant an honorary degree of Master of Arts from Cambridge and was made a member of Trinity College.[1] The election was described more than twenty years later by the Reverend William Cole, who was a student at Cambridge in 1739:

> Mr. Colson was vicar of Chalke, near Gravesend. I think he was of neither University: a plain, honest man, of great industry and assiduity; but the University was much disappointed in their expectations of a Plumian Professor that was to give credit to it by his lectures. He was opposed by old Mr. *De Moivre*, who was brought down to Cambridge, and created M.A. when he was almost fit for his coffin: he was a mere skeleton, nothing but skin and bones, and looked wretchedly, not unlike his mezzotinto print which I have of him. Mr. Colson died at Cambridge, Jan. 1760, rector of Lockington, in Yorkshire.[2]

W. W. Rouse Ball used this anecdote, in part, for his description of the election in *A History of the Study of Mathematics at Cambridge.*[3] Rouse Ball ignored that Cole had erred in Colson's education and in saying it was the Plumian Professorship. Colson received a Master of Arts degree from Cambridge in 1728.[4] Rouse Ball reduced the physical description of De Moivre to "very old and almost in his dotage." He added that Robert Smith, Master of Trinity College, supported Colson. What both Ball and

Cole left out was that there was a third candidate. He was the Reverend Dr. Roger Long, an astronomer and master of Pembroke College, Cambridge.[5] At the time of the election, both Colson and Long were 59 years old.

Cole's anecdote illustrates something that De Moivre's friend Matthew Maty mentioned in his biography of De Moivre. It was the discrimination that De Moivre had to endure as a foreign-born person living in England, even as a citizen. Cole's description of De Moivre looks as if it is meant to defend the decision to appoint Colson—De Moivre already had one foot in the grave, so Colson was the obvious choice. His description conflicts with Charles-Étienne Jordan's and Jean Des Champs's recollections when they met De Moivre and found him to be very good company.[6] These meetings occurred approximately around the time that De Moivre was considered for the position at Cambridge. Perhaps aging, De Moivre was not exactly decrepit; he was still successfully giving mathematics lessons in 1742. Some of Cole's other prejudices perhaps stare out at you from his portrait. De Moivre was definitely thin. Cole looks like he enjoyed his food—a lot of it.[7]

Although Cambridge was done with De Moivre, De Moivre was not quite done with some of the people at Cambridge, dead or alive. He was involved in a minor way with Saunderson's posthumous *Elements of Algebra* and interacted with Colson, who had been his friend since at least the time of De Moivre's dispute with George Cheyne.

Blinded by smallpox as a one-year-old, Saunderson entered Cambridge in 1707 at the age of twenty-five. Within months, he began lecturing young undergraduates on Newtonian physics as well as mathematical topics taken from Newton's *Arithmetica Universalis*, which came out in 1707. He was very successful as a lecturer. In his early years at Cambridge, Saunderson was in contact with Newton, Halley, De Moivre, and other mathematicians.[8] When William Whiston was ejected from the Lucasian Professorship in 1711 for his religious beliefs, Saunderson replaced him. His candidacy was supported and promoted by two of De Moivre's friends: Isaac Newton and Francis Robartes. Like De Moivre, Saunderson, at the time, had no university degree and so was ineligible for the professorship. With Robartes's help, Saunderson was given the degree of Master of Arts by royal warrant in 1711. Quickly thereafter, he was made Lucasian Professor of Mathematics. It was just prior to this time that De Moivre was looking for an academic position on the Continent and Robartes, recognizing his talents, was encouraging De Moivre to write on probability.

While Saunderson was alive, De Moivre communicated two results to him that Saunderson included in his *Elements of Algebra*.[9] De Moivre's first result is related to the foliate curve that he had studied in 1715. This curve is given by the equation $y^3 + y^2x + yx^2 + x^3 = axy$, where a is a known constant. Rather than this curve, De Moivre considered the curve given by the equation $y^3 + y^2x + yx^2 + x^3 = a$. When he took the sum of squares of the terms on the left side of the equation and set it equal to b, a known constant, this gave him two equations in two unknowns. From the first equation and the second, given by $y^6 + y^4x^2 + y^2x^4 + x^6 = b$, he was able to find closed-form solutions for x and y. The second result is similar to the first. In this case the initial equation is $y^4 + y^3x + y^2x^2 + yx^3 + x^4 = a$. Given the sum of squares of the terms on the left side is equal to b, De Moivre was again able to get closed-form solutions for x and y. After some simplification, he applied these kinds of problems to his teaching. In the notebook of his lessons from 1742,[10] he asked his student to solve for x and y from the equations $y^2 + yx + x^2 = 21$ and $y^4 + y^2x^2 + x^4 = 273$.[11]

On September 26, 1738, about eight months prior to his death, Saunderson wrote to De Moivre about a problem in the arithmetic of complex numbers:

> Pray, when you write next, be so good as to let me know, whether you have any thing by you relating to the extraction of the cube root of an impossible binomial, such as $-5+\sqrt{-2}$, or $-5-\sqrt{-2}$, or whether in your reading you have met with any way of doing this with the same certainty as in the case of a possible binomial: for my part, I have met with nothing to the purpose about it, not even in *Wallis* himself, who attempts it.[12]

It is obvious that Saunderson did not think that De Moivre was in his dotage. De Moivre did have something, but had misplaced it in a heap of papers. When he finally wrote, it was to the editor of Saunderson's book; Saunderson had been dead for nearly a year. De Moivre commented that John Wallis used a circular argument in trying to solve the problem in his *Algebra* about fifty or so years before. De Moivre found a solution for the cube root that required the trisection of an angle. This harks back to his 1707 paper in which he used trigonometric arguments to find the roots of certain polynomials of degree three, five, seven, and so on. It is no surprise that De Moivre went on to write in his letter that his method could be generalized to find $\sqrt[n]{a+\sqrt{-b}}$, the nth root of any complex number. Like 1707, the method involves the use of cosines that depend on the value of n, and in this case a and b as well.

There are two other interesting tidbits about De Moivre's letter in Saunderson's book. First, when pointing out Wallis's circular argument, De Moivre references pages 190 and 191 of Wallis's *Algebra* for this error. This is the Latin, not the English, edition of Wallis's work. If he bought the book soon after he arrived in England, he would have been more comfortable reading it in Latin. The second tidbit is that his solution for $\sqrt[n]{a+\sqrt{-b}}$ was presented to the Royal Society at a meeting of January 8, 1740, nearly four months before he sent his letter to the editor of Saunderson's book. The minutes of meeting duly record that "a Paper communicated from Mʳ De Moivre was shewn to the Society."[13] After a brief description of the problem, the entry concludes with "Mʳ De Moivre was ordered thanks for his curious Communication." Full stop. In other words, it was read at the meeting but not recommended for publication in *Philosophical Transactions*. Very few mathematical papers appeared in *Philosophical Transactions* at this time.

De Moivre's interaction with his old friend John Colson in 1742 has its origins earlier in the century. One of the great problems of navigation was the determination of longitude at sea. The traditional method was to measure, from a known fixed point, the time of departure, as well as the speed and direction of the ship. Finding an accurate speed requires having an accurate clock. There were two ways to do this: build a clock that would operate accurately at sea, or use astronomical observations that could be obtained accurately at sea. For example, in the 1680s Edmond Halley considered a method of determining longitude that relies on astronomical measurements obtained by observing the disappearance of fixed stars as the moon passes in front of them. Using new astronomical observations obtained at the Greenwich Observatory, he promoted this method in a paper in *Philosophical Transactions* in 1731.[14]

Halley's proposal was one of the responses to a 1714 Act of Parliament that established a Board of Longitude to adjudicate a prize to be offered to anyone who could measure longitude at sea accurately. A prize of £10,000 was offered to anyone who could determine longitude to within one degree, £15,000 for two-thirds of a degree, and £20,000 for one-half of a degree. Anyone showing promise could be reimbursed up to £2,000 in order to bring the invention up to working order.

The clockmaker John Harrison began working on a special pendulum clock in 1726 that could keep accurate time at sea. Ten years later, Harrison successfully constructed and tested a clock that improved longitude reckoning greatly. He was awarded £500 to continue developing his clock. By 1741, Harrison was working on a second improved clock. It was examined by a group of twelve members of the Royal Society, which included Abraham De Moivre. Also in the group were three of De Moivre's longstanding friends (John Colson, Edmond Halley, and William Jones) and three of De Moivre's former students and friends (Martin Folkes, George Parker, 2nd Earl of Macclesfield, and Lord Charles Cavendish). These members of the Royal Society were so impressed by Harrison's work that in January 1742 they wrote as a group to the Board of Longitude asking the board to continue Harrison's funding.[15]

Harrison's invention was successful, but he did not receive his prize until 1763. And it was for the lowest amount—£10,000. The story of Harrison, the problems he faced, and the discrimination against him has been popularized in Dava Sobel's book *Longitude*, which was made into a television series in 2000 with British actor Michael Gambon playing the role of John Harrison.

De Moivre had at least one other brush with the problem of the determination of longitude at sea. The only woman to try for the Board of Longitude prize, Jane Squire, proposed an astronomical solution, rather than a mechanical clock. Her proposal was to divide the sky into more than a million numbered spaces, which she called "cloves." Based on the clove directly above the navigator at sea, and using an astral watch that was set to the movement of the stars, the navigator could calculate the longitude from Squire's prime meridian, which ran through the alleged spot of Jesus of Nazareth's manger at Bethlehem. Her proposal was not taken seriously, so in response she published in 1742 her correspondence with the commissioners and other scientists.[16] De Moivre was one of her correspondents. He was very kindly and patient, but pointed out one of the flaws in her method. The direction and distance travelled by a ship could not be accurately measured in practice.

In his twilight years, De Moivre remained active in the Royal Society but may not have attended many meetings. The last paper that he wrote was on a topic in life annuities. William Jones presented it to a meeting of the Royal Society in 1744 and it was subsequently published in *Philosophical Transactions*.[17] The problem that De Moivre tackled is how to evaluate an annuity in which there are annual payments made up to the year of death and a final payment proportional to the time elapsed between the last annual payment and the time of death. He may have attended a meeting about two years later on November 13, 1746, when a letter to him from the composer Johann Christoph Pepusch was presented to the Royal Society. It was subsequently published.[18] Pepusch was trying to reconstruct the different types of ancient Greek music whose theory, originating with Pythagoras, was based on ratios of string lengths corresponding to the prime numbers 2, 3, and 5, he said.[19] He added that other ancient writers such as Ptolemy added the prime numbers 7 and 11. Since the theory was mathematically based, Pepusch sought De Moivre's help. De Moivre and his former student, George Lewis Scott, went over the ancient texts on music to try to explain their mathematical content to Pepusch. Pepusch's Latin was weak and his Greek even weaker, which caused De Moivre in frustration to refer to Pepusch as "a stupid German dog, who could neither count four, nor understand anyone that did."[20] Probably prior to De Moivre's frustration setting in, De Moivre and Scott, along with a few others, sponsored Pepusch's nomination to fellowship in the Royal Society. He was put up because he had "distinguished himself by his curious enquiries into the Theory and antiquities of the Science of Musick."[21] De Moivre's other main activities in the Royal Society around this time appear to be sponsoring candidates for election to fellowship. In the decade covering 1737 to 1747, De Moivre was one of the sponsors on fifteen successful nominations for fellowship.[22] At least four were former students. After a six-year hiatus, in 1753 De Moivre signed the nomination papers for one more candidate for fellow. This was Robert Symmer, who was a former student of Colin Maclaurin at Edinburgh.[23]

Although he never received a patronage position, academic or otherwise, that would reflect his talents, De Moivre's abilities were publicly recognized late in his life. This was done, of course, in a non-monetary way in line with the negative view on the British support for science held by the French historian and art critic, Jean-Bernard Le Blanc.[24]

In 1735, when he was in his late sixties, De Moivre was made a member of the Berlin-Brandenburgische Sozietät der Wissenschaften, known in England as the Berlin Academy of Sciences. It came about through the sponsorship of Philippe Naudé. A fellow Huguenot whose family escaped to Berlin instead of London after the Revocation of the Edict of Nantes, Naudé had read De Moivre's *Miscellanea Analytica* and was impressed enough by the work to put De Moivre's name forward to the Berlin Academy. Two years after his election in Berlin, De Moivre returned the favor by acting as one of Naudé's sponsors for fellowship in the Royal Society.[25]

Another non-Brit recognized De Moivre's talents in a different way. Born in Geneva, the medalist Jacques-Antoine Dassier had studied in France and Italy. He came to England in 1740, where he took a position as an assistant engraver at the Mint within a year of his arrival. Prior to taking his position at the Mint, he proposed to make a set of thirteen medals featuring the busts of distinguished Englishmen who were still living. The medals would be financed by subscription at a cost of four guineas for the set or seven shillings sixpence for an individual medal. De Moivre's medal was the second or third one made. His medal and that of the poet Alexander Pope were cast in 1741. The medal set also features two of De Moivre's former students: Martin Folkes and John Montagu, 2nd Duke of Montagu. Folkes's medal, the first of the set, was cast in 1740 and Montagu's in 1751, about two years after his death.[26]

De Moivre also went from a medal to a statue. At some point in the eighteenth century, his bust, along with one of Newton and another of Pope, stood in the orchestra pavilion at Vauxhall Gardens, a pleasure garden that was very near the Thames River in South Lambeth.[27]

Finally, De Moivre had some hope of a monetary reward. On the death of the Prussian philosopher and mathematician Christian Wolf, a position as *associé étranger* became vacant in the French Académie royale des sciences. There were a fixed number of foreign members in the Académie royale and a new member was elected only on the death of a current member. The standard procedure was that the members of the Académie royale would put forward two or more nominations to the king, who would select one. After Wolf's death, the members met on August 14, 1754, and selected De Moivre and the Swiss biologist Albrecht von Haller for the king's consideration.[28] The king chose De Moivre. According to Maty, De Moivre considered his election the crowning achievement of his career.[29] Within a week, the news reached London.[30] After reporting his election and commenting that he had a long and distinguished career, the London newspapers added, "It is also said, that his Most Christian Majesty is inclined to bestow a Pension on him, as a Mark of his own Esteem for Science." The inclination did not have to last long; De Moivre was dead within a little more than three months. De Moivre was replaced by his former student, George Parker, 2nd Earl of Macclesfield. There was no mention of a pension when his election was announced.[31]

By 1744, De Moivre was starting to have problems with his eyesight. In a letter that year to Philip Stanhope, 2nd Earl of Stanhope, De Moivre mentioned that he had wished to reply to Stanhope's questions at greater length, but the weakness of his eyes prevented him from doing so. His handwriting was also beginning to look a little shaky.[32] Over the next decade, his sight and his hearing declined further.[33] In May of 1751, the newspapers reported that he was dangerously ill.[34] De Moivre survived that particular scare. Just prior to his death, he was in serious physical decline, sleeping twenty hours a day. His mind remained sharp, however. Matthew Maty reports,

> Although he came to need twenty hours sleep, he spent the remaining three or four hours taking his only meal of the day and talking with his friends. For the latter, he remained the same: always well-informed on all matters, capable of recalling the tiniest events of his life, and still able to dictate answers to letters and replies to inquiries related to algebra.[35]

He died on November 27, 1754, and was buried from St. Martin-in-the-Fields Church on December 1.[36]

In his will, which he drew up ten years prior to this death, De Moivre left an estate of £1600 invested in South Sea Annuities.[37] His first priority was to provide for his relatives. Amounts of £600 were bequeathed to both Sarah and Anne De Moivre, the two daughters of his nephew Daniel junior. The yearly interest on £400 in these annuities was given to Anne De Moivre, the widow of his brother Daniel senior. On the death of his sister-in-law, her money was to go to his grand-nieces. Daniel junior's widow, Marianne Gomm, had remarried and she was to receive only the annuity payments for her daughters until they reached the age of twenty-one or were married. In the event that his grand-nieces both died before the age of twenty-one, an event which did not occur, £1400 of his investments was to be divided among the survivors of his friends and one relative: Martin Folkes, Edward Montagu, William Jones, Peter Wyche, and John Le Sage. The remaining £200 would be split evenly between the French Hospital and the poor of the Parish of St. Anne, the parish in which he resided. Excluding his cousin Le Sage, the remaining four beneficiaries were all fellows of the Royal Society. Of these, all were fairly wealthy landowners except Jones. De Moivre's estate was administered by his executors, George Lewis Scott and Francis Philip Duval. Scott, a barrister, and Duval, a physician, were both fellows of the Royal Society. Both had De Moivre as one of their sponsors for election to the fellowship. De Moivre directed that his books be sold and that Scott be given his manuscripts with the condition that they would not be published. Not only were the manuscripts never published, none of Scott's own manuscripts or any in his possession seem to have survived.

Two codicils were attached to the will. The first, dated March 9, 1751, instructed his executors to give John Gray the sum of £5 as a mark of his esteem for the man. Gray was elected a fellow of the Royal Society in 1732 and was a friend of James Stirling.[38] Halley, Stirling, and Machin were Gray's sponsors for his fellowship. De Moivre did not know Gray well prior to 1744 when he made out his will. According to De Moivre, Gray had been out of the country for some time; according to Gray's 1769 will, he owned a sugar plantation in Jamaica, which may account for his absence.[39] When he returned to England, he lived on a street by Covent Garden near where De Moivre lived. When Robert Symmer was nominated for fellowship in 1753, Gray was one of his sponsors along with De Moivre. The second codicil is dated August 2, 1752. In it De Moivre instructed his executors to give Susanna Spella £25. She had been his housekeeper for several years; and the only condition to the bequest was that she had to be his housekeeper at the time of his death.

In the decade between his correspondence with Stanhope in 1744 and his death in 1754, despite failing eyesight and other problems that are attendant on old age, De Moivre continued to work as a consultant in the valuation of annuities and games of chance. He had help from his former student, James Dodson. This information surfaces only after De Moivre's death. Within two weeks of De Moivre's death, an advertisement was placed in the London Evening Post that confirms his ongoing consulting work in annuities and games of chance.[40] An unnamed individual in the advertisement said that he had been helping De Moivre in his consulting work for several years because of De Moivre's impaired eyesight. That unnamed individual, so the advertisement said, was now going to continue De Moivre's consulting business at Pons Coffeehouse, one that De Moivre frequented with his Huguenot friends.[41] What points to Dodson as the unnamed individual is that about nine months after the initial notice to the public about the consultant replacing De Moivre, Dodson placed an advertisement in the same newspaper stating that he was in the business of carrying out surveys of estates in land. He could be found at Pons Coffeehouse, at Bank Coffeehouse, or at the Royal Mathematical School.[42] Between the time of De Moivre's death in November 1754 and the placing of the Dodson's advertisement in September 1755, Dodson obtained the position of master at the Royal Mathematical School.[43]

In a letter to Georges-Louis Leclerc, Comte de Buffon, written in the late 1730s or early 1740s, Jean-Bernard Le Blanc confirmed that De Moivre gave advice to gamesters on the calculation of probabilities. In his letter he says,

I must add that the great gamesters of this country, who are not usually great geometricians, have a custom of consulting those who are reputed able calculators upon games of hazard. M. de Moivre gives opinions of this sort every day at Slaughter's coffee-house, as some physicians give their advice upon diseases at several other coffee-houses about London.[44]

By the time this was filtered through Victorian sensibilities, Samuel Smiles transformed the information in this quotation to "It is said he derived a precarious subsistence from fees paid to him for solving questions relative to games of chance and other matters connected with the value of probabilities."[45]

The only thing precarious about De Moivre was perhaps his health; financially he was very comfortable. Le Blanc also indicated the kind of help that gamesters might receive from mathematicians. Professional gamesters often wrote out tables of probabilities giving the various chances in each of the games they played. They either had the table memorized or kept it in their pockets when at the gaming table.

Going to see De Moivre to get a probability table constructed was a convenient walk for a gambler. Many gaming houses were not far from Slaughter's Coffeehouse; there were several of them in Haymarket and Covent Garden in the 1730s.

De Moivre continued to answer other requests for mathematical advice. Sir Alexander Cuming, who had inspired De Moivre to obtain the normal approximation to binomial probabilities, wrote to De Moivre in 1744 from debtors' prison about a problem in finding limits. It was a challenge problem with a trick to it that spoke to Berkeley's criticism of Newton's approach to the calculus in *The Analyst*. Cuming also wrote to Colin Maclaurin and got a different answer from him.[46] The problem is to find the sum for infinite series

$$\frac{a(b-a)}{(a+b)^2}\left(1+\frac{b}{a+b}+\left(\frac{b}{a+b}\right)^2+\left(\frac{b}{a+b}\right)^3+\cdots\right)$$

when $a = 0$. The problem highlights Berkeley's point that at the beginning of a problem a term is nonzero and then later is set to zero. The term within the large round brackets is an infinite geometric series which sums to $(a + b)/a$. Consequently, the complete expression reduces to $(b - a)/(a + b)$ when a is nonzero. When a is then set to zero in this final expression, the sum reduces to 1. An answer of 1 was De Moivre's response to Cuming's query. Maclaurin noted that when the infinite series is truncated to any finite number of terms, the sum is 0 when $a = 0$. On taking the limit by infinitely increasing the number of terms, he decided that 0 was the correct answer, even though he recognized that the reduced expression $(b - a)/(a + b) = 1$ when $a = 0$. The theory of limits was not well understood as the calculus was developing in the seventeenth and eighteenth centuries.[47] After obtaining different answers from De Moivre and Maclaurin, Cuming wrote to Philip Stanhope about it, challenging him for a solution.[48]

The bookseller, Andrew Millar continued to advertise the second edition of *Doctrine of Chances* along with *Miscellanea Analytica* and the fourth edition of *Annuities on Lives* until a third edition of *Doctrine of Chances* was advertised for sale on January 24, 1756.[49] Until at least 1760, Millar advertised the sale of the new edition as well as of the other two books.[50] He continued the book as a luxury item. The price of the new edition was the same as the second, one guinea. It was bound and printed on Royal paper, which was heavy good quality paper, the second most expensive available.[51]

Just as James Dodson helped De Moivre in his consulting work, someone else helped De Moivre bring the third edition of *Doctrine of Chances* to press. This was Patrick Murdoch, a Church of England clergyman who was also a good mathematician. A contemporary and fellow student with Robert Symmer at Edinburgh, he had also studied under Maclaurin. One possible reason De Moivre chose Murdoch was that the clergyman had gone through Maclaurin's manuscripts after his death in 1746 and put together the book *An Account of Sir Isaac Newton's Philosophical Discoveries, in Four Books* that was published in 1748.[52] The book "ranks as one of the most adept popular expositions of Newtonian natural philosophy published in the Enlightenment."[53] For the third edition of *Doctrine of Chances*, De Moivre took a copy of his second edition and made some marginal notes and corrections in it.[54] Murdoch made a few more corrections and additions, then reordered the material. The probability problems in the new edition are unaltered from the old one, other than some shuffling and renumbering of them. There are no new results; rather some additional explanatory material is added to some of the problems. Murdoch took all the material on annuities and placed it at the end of the book. Almost the entire section on annuities is the same, word-for-word, as the fourth edition of *Annuities on Lives* with the errata corrected. There is an additional chapter in the annuities section. This is Chapter IX in which De Moivre (or perhaps Murdoch or Dodson) admits the problem of the incompatibility of the linear and exponential survivorship models in the evaluation of joint-life annuities. Another approach is given that is claimed to be a more accurate approximation to the true value of the joint-life annuity.

All these changes, mostly minor in substance when compared to De Moivre's original results, make the third edition run to 348 pages. Since the second edition is 259 pages, Millar advertised that the book had been increased by a third. Millar may have wanted De Moivre to complete a third edition because his bookshop was running out of copies of the second one.

There were some changes to third edition that may not have come directly from De Moivre before he died. On March 18, 1755, Murdoch wrote to Philip Stanhope about the changes he hoped to make:

> The Edition which Mr. De Moivre desired me to make of his Chances is now almost printed; and a few things, taken from other parts of his work, are to be subjoined in an Appendix. To which Mr. Stevens, and some other Gentlemen, propose to add some things relating to the same subject; but without naming any author: and he thought if your Lordship was pleased to communicate anything of yours, it would be a favour done the publick. Mr. Scott also tells me, there are in your Lordship's hands two Copy Books containing some propositions on Chances, which De Moivre allowed him to copy. If your Lordship would be pleased to transmit these (to Millar's) with your judgement of them, it might be a great advantage to the Edition.[55]

An analysis of Stanhope's unpublished work in probability shows that it is unlikely that Stanhope submitted any of his work.[56] It is impossible to say what new parts of the third edition are due to Murdoch and his friends. There is one reasonable possibility that can be singled out. In the appendix at the end of the third edition, item III is about Waldegrave's problem or the problem of the pool for four players.

In this addition, the writer shows how to obtain the probability that the pool will be won after a given number of games.[57] Anders Hald was puzzled about why De Moivre waited until the third edition when it could easily have appeared in the first edition of *Doctrine of Chances*.[58] The puzzle might be solved if Murdoch or Henry Stewart Stevens or George Lewis Scott inserted this material.

I have told you of De Moivre's fights with Cheyne, Montmort, and Simpson and how he unilaterally ceased correspondence with the Bernoullis, Johann and Nicolaus. I have also told you about a possible cantankerous streak in De Moivre, backed by a few examples. That was not how his friends remembered him. The notice in the newspaper the day after his death reads,

> Yesterday Morning, died of old Age, Mr. Abraham De Moivre, about eighty-seven. His great Knowledge, his communicative Disposition and chearful [sic] Temper render'd him admired, esteemed, and loved by all who knew him.

The substance of De Moivre's death notice is unusual for mid-eighteenth-century newspapers by speaking to qualities that show De Moivre as a person. By contrast, the 1761 notice of Thomas Bayes's death only says that he died suddenly. For Martin Folkes, who died a few months before De Moivre, the death notice lists his accomplishments only: president of the Royal Society, president of the Society of Antiquaries, graduate of both universities and Doctor of Laws at Oxford. Likewise, another De Moivre student higher up the social ladder received only a few more lines than Folkes. When George Parker, 2nd Earl of Macclesfield, died in 1764, after his list of major duties (teller of the Exchequer, president of the Royal Society, vice president of the Foundling Hospital), the death notice mentions in addition only his advantageous marriage and his heirs.[59] Of these three—Bayes, Folkes, and Macclesfield—there is no indication given in their death notices of any positive personal traits of the individual. De Moivre, on the other hand, had a few admirable ones that people publicly recognized. Such was his personality that his network of friends admired, respected, supported, and remained with him until the very end.

Endnotes

Preface

[1] Galloway (1839, pp. 7, 14–15).
[2] Bellhouse (2004) and Dale (2003) are the major treatments of Bayes.
[3] Bellhouse and Genest (2007).
[4] Maty (1755).
[5] Schneider (1968).
[6] Hald (1990).
[7] De Moivre (1984).
[8] De Moivre (2009).

Chapter 1

[1] Maty (1755, p. 44).
[2] Abraham De Moivre's name is variously spelled "De Moivre," "de Moivre" and "Demoivre." Common French usage is "de" rather than "De." I will use "De Moivre" since that is what appears on most documents that I have seen where his signature is attached. The use of "de" or "De" usually implies that the family is noble, although that is not always the case. Abraham De Moivre came from the bourgeoisie. In his biography of De Moivre in the online *Oxford Dictionary of National Biography*, Ivo Schneider suggests that the "de" was adopted in England to gain prestige with his English clients among landed families and the nobility. This was questioned by Bellhouse and Genest (2007) who note that several (140 out of about 1600) Huguenots, overwhelmingly non-noble, recorded at the Savoy Church on arrival in England used the particle "de" in their names. In a genealogy of the Moivre family in the eighteenth-century manuscript (Ms. 171) held by La Bibliothèque de la Société de l'Histoire du Protestantisme français, the particle "de" is used throughout the discussion of this family. When referring to a family by surname only, when the name contains "de," it is usual to drop the "de." I will follow this convention, except when referring to Abraham. For example, "De Moivre's rival in probability was Montmort (Pierre Rémond de Montmort)." Other eighteenth-century variations in the spelling of De Moivre's name have cropped up. In French sources, I have seen Moyvre instead of Moivre. I have also seen Moavre and Movire in early English sources that might reflect non-French speakers trying to write down what they heard. Finally, two

English spellings from De Moivre's contemporary John Flamsteed are Moyver and Moiver which reflect how the English pronounced his name even when he was alive.

[3] Many of the details of De Moivre's early life are recorded in Maty (1755), which is translated in Bellhouse and Genest (2007). Bellhouse and Genest also collected additional information on De Moivre's life. To this information I have added scraps of new material and some historical background.

[4] The modern designation of its location is in the Department of Marne in the administrative region of Champagne-Ardenne.

[5] Daumas (1969, pp. 494–495).

[6] Tassin (1634).

[7] Konnert (2006, pp. 56–60).

[8] Lavedan et al. (1982, pp. 13–14).

[9] Scoville (1960, pp. 242–244).

[10] Ms. 171, La Bibliothèque de la Société de l'Histoire du Protestantisme français, p. 171.

[11] Hérelle (1908, vol. 3, p. 399).

[12] Barnard (1922, pp. 73–108).

[13] Barnard (1922, p. 104).

[14] Hérelle (1900, vol. 1, pp. 120–131).

[15] Zöckler (1952, vol. 3, p. 40–41).

[16] Hérelle (1900, vol. 1, pp. 131–132).

[17] This is reported in Maty (1755). He says, as translated in Bellhouse and Genest (2007), "Religious zeal, which was not as keen in this city as in the rest of France, did not preclude Catholic and Protestant families from entrusting their children to the same tutors." There is some truth to the statement; however, Protestant families were coming under pressure by the authorities.

[18] Bourchenin (1882, pp. 179–190).

[19] Ibid. (pp. 367, 464–465).

[20] Maty (1755). Translation in Bellhouse and Genest (2007, p. 111).

[21] I have been able to examine the fifth edition, Le Gendre (1668). Le Gendre's work is discussed in Sanford (1936).

[22] See, for example, Tinsley (2001, p. 14).

[23] Maty (1755). Translation in Bellhouse and Genest (2007, p. 111).

[24] Huygens (1657).

[25] See David (1962, pp. 110–122).

[26] Up to about 1640, there was a Mark Duncan at Saumur who taught philosophy and Greek as well as mathematics. It is likely that the Duncan who taught De Moivre was a son of this Mark Duncan. See Bellhouse and Genest (2007, p. 111) for a further discussion.

[27] Des Chene (2002, pp. 183–196).

[28] See Shank (2008) for the history of how Newtonianism eventually prevailed over Cartesianism in France in the eighteenth century.

[29] Claude (1708, p. 53) reports that an edict of 1684 banned Protestants from practicing several professions including that of surgeon. As well, Protestant midwives could not attend childbirths. Douen (1894, p. 223) also reports this with specific reference to Daniel De Moivre.

[30] Hérelle (1905, vol. 1, pp. 48–53, 281–288).

[31] Sturdy (1995, pp. 378–380).

[32] Maty (1755, pp. 6–7).

[33] Haag and Haag (1846–1859, vol. 7, p. 260).

[34] Barthélemy (1861, vol. 2, p. 292).

[35] Jaquelot (1712, pp. 11–12, 61–63).

[36] Douen (1894, pp. 223–224). There is much confusion over the Prieuré de Saint-Martin and who was held there. Haag and Haag (1846–1859, vol. 7, p. 433) state that it was Abraham De Moivre held in the prieuré. As Douen points out, the release date is after the time that the De Moivre brothers arrived in England. He claims that Haag and Haag have mixed up the father and son.

[37] Ewles-Bergeron (1997); translation by Christian Genest in Bellhouse and Genest (2007, p. 113).

[38] Here is one example of an inaccuracy: The full anecdote makes a reference to Abraham De Moivre's daughter. It would have been his niece.
[39] Ms. 171, La Bibliothèque de la Société de l'Histoire du Protestantisme français, p. 171.
[40] Huguenot Society (1914).
[41] Huguenot Society Library, from the manuscript entitled "Conversions et reconnoissances faites à l'Église de la Savoye 1684–1702." Reproduced with permission from the Huguenot Society of Great Britain and Ireland.

Chapter 2

[1] https://republicofletters.stanford.edu/
[2] Porter (1991, p. 21).
[3] Hérelle (1908, vol. 3, p. 401) lists a Morel family, but makes no mention of Judic.
[4] Public Record Office PROB/11/811. Will of Abraham De Moivre. The genealogy of the Moivre family found in Ms 171 from La Bibliothèque de la Société de l'Histoire du Protestantisme français shows a Marguerite Morel as a great-grandmother of Abraham De Moivre.
[5] Fallon (1972, p. 281).
[6] Matthews (1974, pp. 267–268).
[7] Agnew (1874, p. 44).
[8] Murdoch (1985, p. 89).
[9] Nichols (1812, p. 578).
[10] Gwynn (1985, p. 82). Dissenting academies were non-Anglican establishments set up typically for the training of dissenting or nonconformist ministers as well as laypersons.
[11] Yonge (1740, p. 5).
[12] Gwynn (1985, p. 71).
[13] Public Record Office PROB/11/811. Will of Abraham De Moivre.
[14] Public Record Office PROB/11/661. Will of Daniel De Moivre (senior).
[15] Westminster Council Archives. St. Martin-in-the-Fields parish registers for 1754.
[16] Lalande (1982, p. 27).
[17] Reproduced from a 1971 reproduction of Rocque (1756) published by Harry Margary and Phillimore & Co.
[18] I have checked the Poor Law Rate Books in the Westminster Council Archives for the parishes of St. Anne and St. Martin-in-the-Fields. The rate books for St. Giles-in-the-Fields are unavailable.
[19] Public Record Office PROB/11/811. Will of Abraham De Moivre.
[20] Rocque (1756).
[21] Cowan (2005, pp. 169–180).
[22] Society for the Diffusion of Useful Knowledge (1837, p, 380).
[23] Burney Collection: *Daily Courant*, January 14, 1710.
[24] Burney Collection: *Post Boy*, September 29, 1702.
[25] For example, over the years 1707–1710 notices were placed in the London newspaper *Daily Courant* for a lost dog, a lost watch, and a lost tankard to be taken to Slaughter's Coffeehouse, if found. *Daily Courant* August, 19 1707, May 4, 1708, and May 29, 1710. See Burney Collection.
[26] Harvey and Grist (2006, p. 164).
[27] Burney Collection: *Daily Post and General Advertiser*, February 24, 1744; *General Advertiser*, October 23, 1745; and *London Evening Post*, May 23, 1749.
[28] Lillywhite (1963, p. 530).
[29] Twiss (1787, pp. 121–122, 163).
[30] Ozanam (1725, Vol. 1, pp. 266–269).
[31] St. John's College Library, Cambridge. TaylorB/E4. Letters dated January 25, 1718, and February 29, 1718.
[32] Crace Collection, Department of Prints and Drawings, British Museum.
[33] Hunter (1994, p. 230).
[34] Bellhouse et al. (2009).

35 Flamsteed (1995–2002, vol. 2, p. 1011).

36 As translated by David Joyce at http://aleph0.clarku.edu/~djoyce/java/elements/bookI/bookI.html #defs.

37 Euclid (1570); spelling modernized.

38 Dechales (1685).

39 Edwards (1979, pp. 191–192).

40 Edwards (1979, pp. 178–180).

41 Stigler (1999, p. 211).

42 Ferraro (2008, pp. 55–56) and Edwards (1979, pp. 179–181).

43 Berkeley's criticisms of the calculus are in Berkeley (1734) and the various responses to it in the eighteenth century are described in Guicciardini (1989) and Jesseph (1993).

44 Ferraro (2008, p 19).

45 Terrall (2002, p. 43) translated from the original letter in French published in *Revue d'Histoire litteraire de la France*, 1908, Tome 15, pp. 111–112.

46 Ferraro (2008).

47 De Moivre (1718, p. viii).

48 Harris (1710).

49 Royal Society, Cl.P/1/4.

50 Arbuthnott (1710).

51 Cohen (1985, p. 164).

52 Guicciardini (1999, pp. 190–193).

53 Guicciardini (1999, chapter 8).

54 Shank (2008).

55 McGrath (2010, p. 29).

Chapter 3

1 The date given by Maty (1755) is 1687. Bellhouse et al. (2009) have argued for the 1689 date based on when Newton was in London for the Convention Parliament. There is no evidence that Newton was in London in 1687. Further, there is evidence on the Devonshire side to support a 1689 date. Bickley (1911, pp. 170–172) puts Devonshire in Yorkshire rebuilding Chatsworth in 1687. He spent his time in the north plotting against James II and did not arrive in London until late in 1688.

2 Matthews (1974, pp. 267–268).

3 Garnier (1900, p. 14).

4 Maty (1755) lists "Cavendish" as one of De Moivre's students, which probably refers to the younger son James. Had Maty meant William, 2nd Duke of Devonshire, he would have written "Devonshire." Both sons, William and James, were subscribers to *Miscellanea Analytica*.

5 Public Record Office PROB 11/811.

6 As translated in Bellhouse and Genest (2007, pp. 113–114).

7 Cooper (1862, p. 50).

8 Statt (1990, p. 46).

9 Huguenot Society (1923) and Royal Commission on Historical Manuscripts (1910).

10 Arnstein (1993, p. 26).

11 Royal Commission on Historical Manuscripts (1912).

12 Tilmouth (1957).

13 See Miriam Yareni's article in Vigne and Littleton (2001, pp. 404–411).

14 Bellhouse et al. (2009).

15 Murdoch (1992).

16 Murdoch (1992) and Bellhouse et al. (2009).

17 Public Record Office PROB 11/588.

18 Tweedie (1922, p. 192).

19 Climenson (1906, p. 286).

20 Anonymous (1949, p. 103).

[21] Jordan (1735, pp. 147 and 174).

[22] Janssens and Schillings (2006, p. 104) and Janssens-Knorsch (1990, p. 34).

[23] Bellhouse and Genest (2007).

[24] Harvey and Grist (2006).

[25] Motteux (1740, p. 114).

[26] Maty (1755) says they met in 1692. Fatio in an article in *Gentleman's Magazine*, volume 7, 1737, says that he gave some mathematical results to De Moivre before 1692.

[27] Senebier (1786, p. 155).

[28] Maty (1755). See also Bellhouse and Genest (2007, p. 114).

[29] *Histoire des ouvrages des sçavans*, vol. 13, 1697, pp. 452–457.

[30] Leibniz (1962, p. 603).

[31] Maty (1755) and Bellhouse and Genest (2007).

[32] Cook (1993).

[33] Cook (1998).

[34] The division is from a set of lectures published in Latin by Isaac Barrow. The Latin version is Barrow (1685); and English translation is in Barrow (1734).

[35] The definitions given here are from the article "Mathematics" in Ephraim Chambers's *Cyclopædia: or, an universal dictionary of arts and sciences* (Chambers, 1778–88).

[36] Aguilón (1613).

[37] Halley (1696).

[38] Halley (1700).

[39] Sherwin (1706).

[40] De Moivre's letter to Halley outlining the approximation is no longer extant and exists undated in printed form only. It is presumed that the letter predates 1706. The letter was first published in 1761 in an obscure and short-lived mathematics journal. See a series of articles by Archibald (1929), White and Lidstone (1930), and Archibald (1945). Archibald (1929) contains a reprint of De Moivre's letter.

[41] Memorandum from De Moivre to John Conduitt in The Joseph Halle Schaffner Collection, University of Chicago.

[42] Royal Society Journal Book, vol. 9. A thorough discussion in German of much of De Moivre's mathematical work is in Schneider (1968).

[43] De Moivre (1695). An English translation of the paper is in Halley (1706, pp. 128–139).

[44] Newton (1687, pp. 250–253). An English translation with notes is in Newton (1999, pp. 646–649).

[45] Newton (1959–1971, vol. 4, p. 183).

[46] De Moivre (1697).

[47] Royal Society Journal Book, vol. 9.

[48] De Moivre (1698).

[49] Wallis (1699).

[50] Royal Society EL/W2/69.

[51] Mazzone and Roero (1997, pp. 27–28).

[52] British Library, Add. 4284/84-85.

[53] The young friend and the approximation are mentioned in a letter of July 8, 1706, from De Moivre to Bernoulli (Wollenschläger, 1933). Hermann mentions the approximation in a letter of August 26, 1706, to Leibniz (Leibniz, 1962).

[54] Maty (1755). The translation is from Bellhouse and Genest (2007, p. 114).

[55] Gaiter and Horns (1940, pp. 116–117).

[56] Halley (1693).

[57] London Metropolitan Archives MS 12 81/12.

[58] Flamsteed (1997, p. 914). Either Joseph Highmore, the artist who painted De Moivre's only known portrait, was kind to his sitter or De Moivre's case was not severe; the portrait does not show any ravages of the disease. I am grateful to Joanna Hopkins, picture curator of the Royal Society, who examined the original portrait for me.

[59] A brief biography is found in Lasocki (1989).
[60] The flute that is shown was made by the prominent London instrument maker Peter Bressan around 1700. The picture is reproduced courtesy of the Victoria and Albert Museum, London.
[61] Lasocki (1997).
[62] Burney Collection: *Daily Courant*, February 27, 1717.

Chapter 4

[1] Guerrini (1986).
[2] There are at least two primary sources for information on the dispute between Cheyne and De Moivre as well as events leading up to it. These may be found in Flamsteed (1997–2002, vol. 2, pp. 914, 936, 950, 953, 987–988, 991, 1011, 1021), Flamsteed (1997–2002, vol. 2, pp. 8, 11, 47, 70), and Gregory (1937, pp. 15, 19, 20, 39, 43–45). Derek Whiteside has given a description of the dispute that was written before the Flamsteed correspondence was published. His description is in Newton (1981, pp. 14–21). There is also an account of the episode in Guerrini (2000, pp. 68–71), again without reference to the Flamsteed correspondence.
[3] Cheyne (1703).
[4] The letter, quoted in Schneider (1968), appears in Bernoulli (1992, p. 383). The phrase in the original French is: "Ce Jeune écossois m'a dit que M. Newton en est prequ'aux deux tiers de la réimpression de ses principes; & que c'etoit lui qui avoit suscité M. Moivre contre M. Cheynée; & que ces deux Antagonistes sont presentement bien reconciliés & fort bons amis: Dieu en soit loüé."
[5] This is Flamsteed's comment. De Moivre made the same claim three years later in a 1705 letter to Johann Bernoulli (Wollenschläger, 1933, p. 188).
[6] The recollection is quoted in Newton (1981, pp. 16).
[7] Royal Society Journal Book, vol. 9, February 1702/3.
[8] Reported in, for example, Guerrini (2000, p. 70).
[9] Mentioned in a letter from Bernoulli to De Moivre, November 15, 1704; Wollenschläger (1933).
[10] Wollenschläger (1933, p. 181).
[11] Flamsteed (1997, pp. 987–988).
[12] Flamsteed (2002, p. xlvii).
[13] Royal Society Journal Book, vol. 9.
[14] Flamsteed (1997, p. 948).
[15] De Moivre (1704).
[16] Flamsteed's copy is currently held in the British Library, shelf mark 60.a.3.
[17] Royal Society Journal Book, vol. 9. Whoever took the notes for the meeting had a brief lapse and used the English "and" instead of the Latin "et" in the title of the paper.
[18] An alternative approach is to use the results in De Moivre (1698) or the inversion of a series.
[19] In the preface of De Moivre (1704), De Moivre uses the Latin word "amicissimi" or "best friend" to describe Colson.
[20] Newton and Colson (1736, pp. 171 and 310).
[21] All the extant correspondence between De Moivre and Bernoulli is published in Wollenschläger (1933).
[22] Grafton (2009).
[23] Newton (1959–1971,vol. 4, p. 290).
[24] A letter of July 6, 1708 from De Moivre to Bernoulli (Wollenschläger, 1933, p. 253) mentions a recent English victory over the French. The Battle of Oudenarde took place on June 30, 1708.
[25] Cheyne (1705).
[26] Wollenschläger (1933, p. 194). The translation is by Catherine Cox.
[27] Gregory (1937, p. 39).
[28] Maty (1755); Bellhouse and Genest (2007, p. 117).
[29] Bernoulli (1992, pp. 162, 174, 178, 228, and 229).
[30] Wollenschläger (1933, p. 233).
[31] Timperley (1839, pp. 664 and 811).
[32] Bernoulli (1992, p. 383).

33 Cheyne (1724, pp. vi–vii).
34 Newton (1959–1971, vol. VIII, pp. xiv–xv).
35 Wollenschläger (1933, p. 212).
36 Mazzone and Roero (1997, pp. 261 and 264).
37 Wollenschläger (1933, p. 254), as translated by Catherine Cox.
38 Wollenschläger (1933, pp. 213–214).
39 Wollenschläger (1933, p. 224).
40 The substance of the letter is discussed in Guicciardini (1995).
41 Bernoulli (1710, pp. 529–530).
42 British Library British Library, Add. 4284.
43 Perks (1706).
44 Wollenschläger (1933, pp. 220 and 226).
45 Pepusch (1746).
46 Burney (1789, vol. IV, p. 638).
47 Columbia University Library, Smith Collection. The letter is printed in Smith (1922). The Robins in question was Benjamin Robins. The online *Oxford Dictionary of National Biography* for Robins by Brett D. Steele states that Robins found "employment as a mathematics tutor to prospective Cambridge students," which would have put him in direct competition with De Moivre for these students. Smith dates the letter to 1723 or 1724. This dating is almost certainly incorrect. Robins, mentioned in the letter, was born in 1707. Not given in the quotation but mentioned in the same letter is Philip Stanhope, 2nd Earl Stanhope, whom Robins was tutoring in mathematics. Stanhope was born in 1714, and since his uncle and guardian was opposed to the study of mathematics, it would have been at least 1730 or perhaps as late as 1735 when the letter was written.
48 Flamsteed (1997–2002, vol. 2, pp. 987 and 1011).
49 Carnegie Mellon University, Posner Family Collection 523.6 H18A.
50 The identification of Sprat is in Bellhouse et al. (2009).
51 The rule of alligation is defined by the Oxford English Dictionary as "the arithmetical method of solving questions concerning the mixing of articles of different qualities or values."
52 Royal Society Journal Book, vol. 10, March 6, 1705/6.
53 Royal Society Journal Book, vol. 10, November 13, 1706.
54 Lamy (1701).
55 Royal Society Letter Book vol. 14: Correspondence of John Shuttleworth.
56 Shuttleworth (1709).
57 London Metropolitan Archives. Burial register of St. Giles-in-the-Fields. An entry for October 7, 1707 reads "Ann of [ ] Movire." From the examination of other entries, what should appear in the [blank] spot is the husband's name. There were other entries with blanks beside the woman's name so that the blank might signify a widow or the inability of the parish clerk to learn the name of the husband if he was not present at the burial. "Movire" is one of many variant spellings of Moivre in early eighteenth-century records in England.
58 Register of the Church of West Street. Publications of the Huguenot Society (1929, p. 14). Abraham De Moivre stood as godfather to his nephew. Other children of Daniel De Moivre followed—Anne in 1708 and Elizabeth in 1709.

Chapter 5

1 Translation of Maty (1755) from Bellhouse and Genest (2007, p. 130).
2 De Moivre appears among "Persons of other Nations" in Chamberlayne (1710) and Miège (1711) and among the regular members in Chamberlayne (1716) and Miège (1715). A June 26, 1711, entry in the Royal Society Council Minutes lists several people from whom the Council wanted to request donations in aid of new lodgings for the Society. The first sixty names are in alphabetical order and correspond to regular members; the last four names, ending with De Moivre, are also in alphabetical order and correspond to foreign members.

[3] The general integration of Huguenots into English society and the English population's reaction to the influx of immigrants is discussed by Eileen Barrett: "Huguenot integration in late 17th- and 18th-century London: insights from the records of the French Church and some relief agencies." In Vigne and Littleton (2001, pp. 375–382).

[4] Le Blanc (1747, vol. I, pp. 168–169).

[5] Wollenschläger (1933, pp. 240–241).

[6] Leibniz (1962, pp. 844 and 846–849). The whole series of letters are described in Bellhouse and Genest (2007, p. 119).

[7] De Moivre (1707).

[8] De Moivre (1722b).

[9] Schneider (1968).

[10] De Moivre (1718, pp. 29–31).

[11] Wollenschläger (1933, pp. 241–250).

[12] For biographical information on Francis Robartes (1650–1718) see his entry in *Oxford Dictionary of National Biography* by G. I. McGrath.

[13] Hall (1980, Chapter 9).

[14] A discussion of the various eighteenth-century solutions to the gambler's ruin and the duration of play problem can be found in Hald (1990, pp. 347–374).

[15] Royal Society Archives Cl.P/1/4 and Royal Society Journal Book, vol. 8, March 6, 1691/2.

[16] Bernoulli's problem is described in Hald (1990, pp. 184–185).

[17] Cited as Roberts (1693).

[18] Todhunter (1965, pp. 53–54).

[19] Harris (1710).

[20] Montmort (1708).

[21] Newton (1959–1971, vol. 4, pp. 533–534) and Rigaud (1965, p. 256). In a 1712 letter from Abraham De Moivre to Johann Bernoulli, De Moivre says that Robartes had mentioned Montmort's book to him (Wollenschläger, 1933, p. 272).

[22] Anonymous (1709, p. 462). Review of *Essay d'analyse sur les jeux de hazard*, by Pierre Rémond de Montmort, *Supplément du Journal des sçavans* (1709): 462.

[23] De Moivre (1718, p. i).

[24] Cotton (1674 and 1709).

[25] Wollenschläger (1933, p. 272).

[26] Royal Society Journal Book, June 21, 1711.

[27] *Journal de Trévoux ou memoires pour server a l'histoire des sciences et des arts*, 1709, pp. 1369–1383.

[28] *Journal de Trévoux ou memoires pour server a l'histoire des sciences et des arts*, 1712, pp. 1452–1367.

[29] Montmort (1713, p. 362).

[30] Osborne (1742).

[31] Harrison (1978).

[32] Sunderland (1881–1883).

[33] British Library, Add. 4281.

[34] Montmort's receipt of *De Mensura Sortis* is also recorded in Montmort (1713, p. xxvii).

[35] Montmort (1713, p. 375).

[36] Bernoulli (1992, pp. 518 and 523).

[37] Universitätsbibliothek Basel Bernoulli papers. L I a 654, Nr. 9 and L I a 654, Nr. 10.

[38] McClintock's translation of De Moivre (1711) in De Moivre (1984, p. 237).

[39] Montmort (1713, pp. 361–370).

[40] Bernoulli (1992, p. 523).

[41] Universitätsbibliothek Basel Bernoulli papers. L I a 654, Nr. 10.

[42] Montmort (1713, pp. 361–370).

[43] Montmort actually used the word pillaged, rather than plagiarized: "il a bien pillé mon livre sans me nommér." Taylor (1793, pp. 97–98).

[44] Shank (2008).

[45] Fontenelle (1719, p. 89).

[46] See Bellhouse and Genest (2007, p. 121) for a full translation of this passage.
[47] Halley (1693).
[48] Royal Society Archives Cl.P/1/4 and Royal Society Journal Book, vol. 8, March 6, 1691/2.
[49] Schneider (1968) and Hald (1990).
[50] Edinburgh University Library, GB 0237 David Gregory Dk.1.2.2 Folio B [18].
[51] Bellhouse and Davison (2009) have analyzed De Moivre's Problems 6 and 7 and have speculated on how De Moivre went about obtaining his numerical solutions to these problems.
[52] Bellhouse and Davison (2009) have discussed the fluxional equation and the attempted infinite series solution.
[53] Hald (1990, p. 203). An explanation of De Moivre's solution can be found in Hald (1990, pp. 203–240) and Thatcher (1957).
[54] Montmort (1713, pp. 248–257).
[55] Piquet is described in Anonymous (1651) and Cotton (1674). The first mention of the variation in play using the pool is in Seymour (1719).
[56] Seymour (1719, pp. 92–93).
[57] Seymour (1719, p. 91).
[58] De Moivre (1718, p. 84).
[59] See Bellhouse (2007a) for information about Charles Waldegrave and his association with Montmort.
[60] St. John's College Library, Cambridge. TaylorB/E7.
[61] Montmort (1713, p. 369).
[62] Montmort (1713, p. 275).
[63] McClintock's translation of De Moivre (1711) in De Moivre (1984, p. 240).
[64] Lewis and Short (1879).
[65] St. John's College Library, TaylorB/E7.
[66] De Moivre's pride in his work was reported by William Jones in a letter to Roger Cotes. See Edleston (1969, p. 208).
[67] Wollenschläger (1933, pp. 270–271, 279).
[68] Hall (1980, p. 6) has pointed out that two of the sources of several priority disputes in past were "the great value attached to personal merit" and "the emphasis on innovation as the creation of an individual talent."
[69] Halley (1715, p. 251). See also Bellhouse and Genest (2007, p. 120).
[70] Ladurie et al. (2001, p. 220).
[71] See, for example, Bellhouse (2008).

Chapter 6

[1] Baker (1756, p. 155).
[2] Newton (1711).
[3] Hall (1980, pp. 169–170) and Newton (1967–1981, vol. II, pp. 206–207).
[4] Newton (1959–1971, vol. V, p. 95).
[5] See Hall (1980) for a full treatment of the dispute and Keill's role in it.
[6] Newton (1967–1981, vol. III, pp. 244–255).
[7] At the end of De Moivre (1702) it is stated that Newton gave De Moivre access to some of his manuscripts on the quadrature of curves.
[8] The table appears on pages 198 to 204 of Newton (1704). Newton had the theologian and natural philosopher Samuel Clarke translate the *Opticks* into Latin (Newton (1706)). Newton's biographer, David Brewster (Brewster (1855, vol. I, p. 248)), claimed, "Demoivre is said to have secured and taken charge of this translation, and to have spared neither time nor trouble in the task." Brewster gives no source for his information and I can find no source material that supports the claim. It is possible that Brewster mixed up the Latin edition with the French edition that De Moivre did have a hand in. The French edition, the second edition, was published in 1722.
[9] Fatio de Duillier (1699, p. 18).
[10] A detailed description of the whole dispute is in Hall (1980).

[11] Wollenschläger (1933, p. 254), as translated by Catherine Cox.

[12] Hall (1980, p. 178).

[13] Royal Society Archives EL/H3/53.

[14] The claim is in a 1714 letter from Bernoulli to Leibniz. An English translation is in Newton (1959–1971, vol. VI, p. 68).

[15] Universitätsbibliothek Basel Bernoulli papers. L I a 654, Nr. 11.

[16] For example, the Royal Society's journal book shows that on June 25, 1713, De Moivre brought his friend Pierre de Magneville to a Royal Society meeting. The journal book records, "Mr. de Moivre desiring one Mr. de Monville might be present was admitted."

[17] Wollenschläger (1933, pp. 270–271).

[18] Wollenschläger (1933, pp. 277–280).

[19] Hall (1980, pp. 195–197).

[20] Wollenschläger (1933). De Moivre expected the book to come off the press by the end of November 1712. It did not appear until June 1713.

[21] Universitätsbibliothek Basel Bernoulli papers. L I a 673, Nr. 1, fol. 5–6.

[22] Universitätsbibliothek Basel Bernoulli papers. L I a 673, Nr. 1, fol. 7–8.

[23] Wollenschläger (1933, pp. 286–289).

[24] Universitätsbibliothek Basel Bernoulli papers. L I a 673, Nr. 1, fol. 9–10. In his letter to Arnold of May 9, 1714, Bernoulli attributes the phrase "creature de Newton" to Arnold.

[25] Universitätsbibliothek Basel Bernoulli papers. L I a 654, Nr. 19*.

[26] Universitätsbibliothek Basel Bernoulli papers. L I a 654, Nr. 13.

[27] Wollenschläger (1933, pp. 289–290) as translated by Catherine Cox.

[28] Bernoulli (1713, p. 127).

[29] See Hall (1980, p. 295).

[30] There was also, unknown at the time to Bernoulli, a serious faux pas on Newton's part. He did not acknowledge in the second edition of the *Principia* that the correction to Proposition 10 in Book II was due to Bernoulli.

[31] Universitätsbibliothek Basel Bernoulli papers. L I a 673, Nr. 1, fol. 13–14.

[32] *Journal literaire* May/June 1713.

[33] *Journal literaire* November/December 1713.

[34] Newton (1959–1971, vol. 6, pp. 80–90).

[35] Newton's direct involvement with Keill comes third-hand based on a letter, no longer extant, from De Moivre to Varignon. Varignon apparently told Jacob Hermann of it, who informed Christian Wolf, who passed the information on to Leibniz. Wolf's letter is translated in Newton (1959–1971, vol. VI, p. 180).

[36] *Journal literaire* July/August,1714.

[37] Newton (1959–1971, vol. 6, p. 114).

[38] Guicciardini (1995) has treated the controversy between Keill and Bernoulli on central forces in detail. The focus here is on De Moivre's potential involvement in the controversy.

[39] Guicciardini (1995, p. 555).

[40] *Journal literaire* 1716.

[41] Keill's article has a misprint and gives the year as 1708.

[42] *Journal literaire* 1719.

[43] Universitätsbibliothek Basel Bernoulli papers. L I a 665, Nr. 11.

[44] Taylor (1715); the *Methodus Incrementorum* has been studied and analyzed extensively by Feigenbaum (1985).

[45] A pendulum consists of a string, wire, or rigid rod with a weight or bob attached to one end. It is suspended from a fixed point and allowed to swing freely or oscillate. It is a simple pendulum if the mass of the suspending material is much less than the bob and the length of the suspending material is much greater than the dimension of the bob. It is a compound pendulum when the mass of the suspending material is not negligible when compared to the bob.

[46] Leibniz (1716) and Bernoulli (1716).

[47] Taylor (1719).

[48] St. John's College Library, Cambridge. TaylorB/E7.

[49] Messbarger (2002, p. 55).
[50] Meli (1999).
[51] The episode in described in Newton (1967–1981, vol. 8, pp. 62–67) and Newton (1959–1971, vol. 6, pp. 285–293, 295–296).
[52] Englesman (1984, p. 74).
[53] Newton (1959–1971, vol. 7, p. 138).
[54] This was October 1718 and the publication date on the book's title page is 1719.
[55] The exchange of letters is in Newton (1959–1971, vol. 7, pp. 2–3, 14–17, and 50–53).
[56] A list of letters between Varignon and De Moivre is given in Schneider (1968).
[57] The dates of the known letters in the De Moivre-Varignon correspondence, along with some of the context, is given in Schneider (1968, pp. 197–201).
[58] Des Maizeaux (1720).
[59] Newton (1959–1971, vol. 7, pp. 42–47, 69–71, 75–81, 218–223). More on the dispute between Newton and Bernoulli circa 1718 to 1722 can be found in Hall (1980, pp. 238–244).
[60] Newton (1959–1971, vol. 6, pp. 457–458, 463).
[61] Newton (1959–1971, vol. 7, pp. 128–129).
[62] Dobbs (1975, p. 22).
[63] Newton (1959–1971, vol. 7, pp. 130–131).
[64] As translated by Catherine Cox.
[65] Newton (1959–1971, vol. 7, pp. 90–91).
[66] The episode can be pieced together from Newton (1959–1971, vol. 7, pp. 119–123, 141–142, 147–148, 152–156, and 214–215).
[67] Newton (1722). The acknowledgment appears in the last paragraph of the preface.
[68] This passage has often been quoted or paraphrased. I have been unable to track down the original source. The earliest reference that I can find is Charles Bossut (1803, p. 388). This is an English translation from the original French published in 1802.
[69] University of Chicago, The Joseph Halle Schaffner Collection.

Chapter 7

[1] Royal Society Journal Book, vol. 11, December 20, 1716.
[2] De Moivre (1717).
[3] Carnegie Mellon University, Posner Family Collection 523.6 H18A.
[4] Royal Society Journal Book, vol. 11, March 20, 1718
[5] De Moivre (1719). A précis of the paper in English appears in *Miscellanea Curiosa*, vol. I, 1749, pp. 24–25.
[6] Keill (1721, pp. 281–284).
[7] De Moivre (1715).
[8] Newton (1704, 1711).
[9] Halley (1715, p. 251).
[10] De Moivre (1730, p. 148).
[11] St. John's College Library, Cambridge. TaylorB/E4.
[12] Montmort (1717).
[13] St. John's College Library, Cambridge. TaylorB/E7.
[14] De Moivre (1730, p. 148).
[15] Deslandes (1737, pp. 264–265).
[16] Deslandes (1713, pp. 42–42). The original Latin is translated by Elizabeth Renouf.
[17] Another possible rendering is "share your knowledge".
[18] Bernoulli (1713).
[19] The letter from Nicolas Bernoulli to De Moivre is quoted in Peiffer (2006).
[20] The letter from De Moivre to Nicolas Bernoulli is quoted in Peiffer (2006).
[21] Universitätsbibliothek Basel Bernoulli papers. L I a 654, Nr. 19*.
[22] Montmort (1713, pp. 361–370).

[23] Royal Society Journal Book, vol. 11, February 11, 1714.

[24] Bernoulli (1714).

[25] De Moivre (1714).

[26] Hald (1990, p. 390).

[27] The challenge is mentioned in a letter from Gabriel Cramer to Jean-Louis Calandrini that is printed in Galiffe (1877).

[28] Universitätsbibliothek Basel Bernoulli papers. L I a 665, Nr. 13*.

[29] Montmort's letters to Taylor mentioning De Moivre are reproduced in Taylor (1793, pp. 88, 90, 92, and 97).

[30] St. John's College Library, Cambridge. TaylorB/E4. Letter dated July 4, 1716.

[31] St. John's College Library, Cambridge. TaylorB/E4. Letter dated October 17, 1717.

[32] St. John's College Library, Cambridge. TaylorB/E7.

[33] Universitätsbibliothek Basel Bernoulli papers. L I a 665, Nr. 7.

[34] Universitätsbibliothek Basel Bernoulli papers. L I a 665, Nr. 11.

[35] Taylor (1793, p. 92).

[36] St. John's College Library, Cambridge. TaylorB/E7.

[37] Royal Society Journal Book, vol. 11, March 22, 1718.

[38] Burney Collection: *Post Man and Historical Account*, March 29, 1718.

Chapter 8

[1] De Moivre (1707, 1722a).

[2] Montmort (1713, pp. 361–369).

[3] De Moivre (1718, p. ix).

[4] Royal Society Journal Book, vol. 12, May 5, 1720.

[5] Royal Society Journal Book, vol. 11, March 22, 1718.

[6] Royal Society, Classified Papers Cl.P.I.43.

[7] De Moivre (1718, pp. 150–151).

[8] De Moivre (1722a).

[9] Hald (1990, pp. 347–349, 433–437) has provided the mathematical details surrounding this problem.

[10] A reconstruction of De Moivre's solution using the tools available to him at the time is given by Schneider (1968, pp. 288–292).

[11] The allegorical images are interpreted in Bellhouse (2008).

[12] *Oxford Dictionary of National Biography*; biography of Joseph Goupy by Sheila O'Connell.

[13] St. John's College Library, Cambridge. TaylorB/E7. The hand drawing of the semicircle in this letter is reproduced by permission of the Master and Fellows of St. John's College, Cambridge.

[14] De Moivre (1718, p. 1).

[15] Maty (1755, p. 26).

[16] Royal Society, Letter Book, vol. 22, letter from De Moivre to Machin dated January 14, 1735/6.

[17] Kahle (1735).

[18] St. John's College Library, Cambridge. TaylorB/E4. Letters dated December 17, 1717, January 25, 1718, and February 19, 1718.

[19] Universitätsbibliothek Basel Bernoulli papers. L I a 665, Nr. 22*.

[20] Universitätsbibliothek Basel Bernoulli papers. L I a 665, Nr. 12.

[21] St. John's College Library, Cambridge. TaylorB/E4. Letter dated February 8, 1719.

[22] Universitätsbibliothek Basel Bernoulli papers. L I a 665, Nr. 15.

[23] St. John's College Library, Cambridge. TaylorB/E7.

[24] Montmort (1713, p. 366–367).

[25] Montmort (1713, pp. 248–257).

[26] De Moivre (1718, pp. 29–31).

[27] The Prussian mathematician Christian Goldbach published a result leading to a similar formula in 1720. It appears in Goldbach (1720). Goldbach's formula is discussed in Bottazzini (1996, pp. 167–168).

[28] This is the book *Analysis per Quantitatum Series, Fluxiones, ac Differentias*.

29 St. John's College Library, Cambridge. TaylorB/E4. Letter dated February 8, 1719.
30 De Moivre (1718, p. 1).
31 De Moivre (1718, pp. 155–157, 174–175) and Montmort (1713, pp. 177–179).
32 St. John's College Library, Cambridge. TaylorB/E4. Letter dated February 8, 1719.
33 De Moivre (1718, p. 159–162).
34 St. John's College Library, Cambridge. TaylorB/E4. Letter dated February 8, 1719.
35 De Moivre (1730, p. 223). What De Moivre said is that Robartes constructed his table more than twenty years before Montmort published his *Essay d'analyse*. Depending on whether De Moivre is referring to the first or the second edition, the table would have been constructed prior to 1688 or 1693. Since Robartes seems to have been active in probability in the early 1690s, the latter date is more likely.
36 The analyses of Basset are in Montmort (1713, pp. 144–156) and De Moivre (1718, pp. 32–39). Faro is treated in Montmort (1713, pp. 77–104) and in De Moivre (1718, pp. 40–44).
37 Saveur's treatment of Basset is described in Todhunter (1865, p. 46). See Hald (1990, pp. 239–240) for a discussion of Bernoulli's treatment of the game.
38 St. John's College Library, Cambridge. TaylorB/E4. Letter dated February 8, 1719.
39 The restrictiveness of the assumption was first noted by Todhunter (1865, p. 152). See also Hald (1990, p. 303).
40 Translated in Bellhouse and Genest (2007).
41 As mentioned in LeBlanc (1747, Vol. I, p. 86), De Moivre had seen the actor and playwright Colley Cibber perform.
42 Cotton (1709, p. 177).
43 Farquhar (1701, p. 18).
44 Centlivre (1705).
45 Boyer (1700).
46 Harvey and Grist (2006, p. 164).
47 Arbuthnot (1692).
48 De Moivre (1718, pp. xiii and 135).
49 Woodcock's will shows that he left an estate worth at least £10,000 in about 1730. Public Record Office PROB11/653.
50 Challis (1992, pp. 391 and 431).
51 Francis (2008 , p. 48).
52 This expression with a discussion of De Moivre's new algebra and its relation to modern probability notation is in Hald (1990, pp. 336–338).
53 His complaints to Brook Taylor about this problem are in a letter dated February 8, 1719, in St. John's College Library, Cambridge. TaylorB/E4.
54 De Moivre (1718, pp. 85–102).
55 De Moivre (1718, pp. 128–133).
56 St. John's College Library, Cambridge. TaylorB/E4. Letter dated February 8, 1719.
57 St. John's College Library, Cambridge. TaylorB/E7.
58 Hald (1990, p. 434).
59 Schneider (1968).
60 De Moivre (1722b).

Chapter 9

1 *Oxford Dictionary of National Biography*; biography of Colley Cibber by Eric Salmon.
2 Le Blanc (1747, vol. 1, p. 62).
3 Cibber (1728, p. 31).
4 Burney Collection: *London Journal*, January 23, 1727.
5 Anonymous (1731).
6 Burney Collection: *The Universal Magazine*, 1749, vol. 5, p. 263.
7 Burney Collection: *Post Man and Historical Account*, March 29, 1718.
8 Plomer (1922, p. 234).

[9] Plomer (1922, pp. 204–205).

[10] Guicciardini (1989, p. 15).

[11] Burney Collection: *Daily Courant*, July 7, 1704.

[12] Osborne (1742), Baker (1756), and Harrison (1978).

[13] St. John's College Library, Cambridge. TaylorB/E4.

[14] Hazen (1969).

[15] Burney Collection: *Daily Journal*, March 15, 1724.

[16] *Oxford Dictionary of National Biography*; biography of James Stirling by Ian Tweddle.

[17] Stirling's copy of *The Doctrine of Chances* is in the collection of Stephen Stigler.

[18] Dobbs's career at Trinity College Dublin is given in Burtchaell and Sadleir (1922, p. 233).

[19] Royal Society Archives, EL/D2/6.

[20] Wellcome Library, MS. 6146/56.

[21] Royal Society Archives, EL/D2/7.

[22] De Moivre (1738, p. 191).

[23] St. John's College Library, Cambridge. TaylorB/E7.

[24] Cambridge University Library, Add. 9547/4/9 and Add. 9547/4/10.

[25] The rough copy is Add. 9547/4/9 and the copy done in a very fine hand is Add. 9547/4/10. The identification of Jones as the person who wrote the rough copy was obtained by comparing the handwriting in the rough copy to other manuscripts in the collection known to have been written by Jones.

[26] Gravell and Miller (1983).

[27] Voorn (1960, p. 537) and Gravell and Miller (1983, p. 228).

[28] Cambridge University Library Add. 9597/8/9, pp. 18 and 19.

[29] Cambridge University Library, Add. 9597/4/14.

[30] Cambridge University Library, Add. 4000.

[31] Newton (1967, p. xxx).

[32] Westfall (1980, pp. 871–873).

[33] De Moivre (1738, p. 235).

[34] De Moivre (1697).

[35] Maty (1755); translation in Bellhouse and Genest (2007, p. 126).

[36] Galiffe (1877, pp. 5–7).

[37] Tweedie (1922, pp. 96–97).

[38] Tweedie (1922, p. 109).

[39] Edwards (1987, p. 116).

[40] Arbuthnot (1710).

[41] Gowing (1983).

[42] Montmort acknowledged receipt of the problem in a letter to Taylor dated January 1, 1718. He told Taylor that he had sent the problem to both Johann and Nicolaus Bernoulli. St. John's College Library, Cambridge. TaylorB/E4.

[43] Tweedie (1922, p. 141).

[44] A translation of what Smith wrote is in Gowing (1983, pp. 69–70).

[45] De Moivre (1707) and De Moivre (1722b).

[46] Cambridge University Library Add. 9497/4/21.

[47] Dodson (1742, pp. i–xx).

Chapter 10

[1] Burney Collection: *General Evening Post*, December 5, 1745.

[2] Barrow (1734). See also the article "Mathematics" in Chambers (1778–88).

[3] Turner (1973).

[4] Bellhouse et al. (2009, pp. 148 and 158).

[5] Wordsworth (1877, Appendix IV).

[6] Waterland (1730).

[7] Whitehouse (2010).

[8] Dr. Williams's Library, I.e.28.
[9] Bellhouse et al. (2009).
[10] Alumni from Cambridge are in Venn and Venn (1922–1954) and alumni from Oxford are in Foster (1968).
[11] See Bellhouse et al. (2009) for a discussion.
[12] Bellhouse (2007b).
[13] Ward (1695a).
[14] Wollenschläger (1933, p. 240).
[15] The notes are in the Landesbibliothek Oldenburg; they have been examined and described by Folkerts (2004).
[16] De Moivre's letters to Stanhope are in the Centre for Kentish Studies U1590/C21.
[17] Dr. Williams's Library, manuscript lecture notes on mathematics L.237 and L.238.
[18] Cambridge University Library, Macclesfield Collection, MS Add. 9597/4/19.
[19] Newton (1707, pp. 81–82).
[20] Ball (1912, p. 330).
[21] It is interesting to note in passing that Newton's statement of the travelling problem was originally in Latin, Jones's in English, and De Moivre's in French.
[22] Le Gendre (1668, pp. 153 – 158).
[23] Cambridge University Library, Macclesfield Collection, MS Add. 9597/8/10–12.
[24] Cambridge University Library, Macclesfield Collection, MS Add. 9597/8/9.
[25] Trevigar (1731).
[26] Bertrand (1874).
[27] I am grateful to my former graduate student Ksenia Bushmeneva who, as a native Russian, carried out the enquiries for me in Russia and Poland.

Chapter 11

[1] Universitätsbibliothek Basel Bernoulli papers. L I a 22, Nr. 180a.
[2] Bernoulli's dissertation entitled *Dissertatio inauguralis mathematico-iuridica de usu artis coniectandi in iure* was published in Basel in 1709. It is unlikely that De Moivre saw this publication. Even today it is not held by the British Library, the Royal Society Library, or any of the libraries at Oxford or Cambridge. The *Acta Eruditorum* would have been more readily available to De Moivre.
[3] De Moivre (1725).
[4] Halley (1693).
[5] Burney Collection: *London Daily Post and General Advertiser*, July 11, 1739.
[6] *Oxford Dictionary of National Biography*; biography of Thomas Parker, 1st Earl of Macclesfield, by A. A. Hanham.
[7] Daston (1988, pp. 168–169).
[8] Richards (1730, pp. 4–7) describes four types of land tenure, but the three given here encompass all the possibilities.
[9] See, for example, Hayes (1726, pp. 84–85).
[10] Evans (1817, pp. 397–398).
[11] See, for example, Recorde (1662) and Wingate (1668).
[12] The associations are established and described in detail in Bellhouse et al (2009).
[13] The letter from De Moivre to Halley on the interest rate calculation is reproduced in Archibald (1929). Other relevant articles are White and Lidstone (1930) and Archibald (1945).
[14] De Vries and van der Woude (1997).
[15] Childs (1991, p. 1).
[16] British History Online: *Statutes of the Realm*, vol. 6, 1685–1694. William and Mary, 1692, c. 3, part vii.
[17] Halley's work is discussed in a number of places. My current favorite reference is Bellhouse (2011).
[18] Royal Society Journal Book, vol. 8.
[19] The errors are discussed in Bellhouse (2011).
[20] Burney Collection: *Post Man and Historical Account*, September 4, 1703.

[21] The two given here are described in Walford (1871, vol. I, pp. 108 and 110). An extensive description of the Mercer's Company scheme is in Clark (1999).

[22] Rogers's advertisements appear in *London Journal* (Burney Collection) beginning in 1724.

[23] The data are taken from Walford (1871).

[24] Burney Collection: *London Gazette*, December 22, 1722.

[25] Cherry (2001). Her biographical description of Hatton conflicts with Taylor (1954, p. 293), who puts Hatton in Stourbridge, Worcestershire. Cherry's reconstruction of Hatton's career is much more convincing than Taylor's brief description.

[26] Ward (1695b).

[27] Ward (1707, pp. 274–275).

[28] Walford (1871, vol. I, p.113).

[29] Ward (1710, pp. 107–113).

[30] Hatton (1714, pp. 113–115).

[31] Hatton (1721).

[32] Burney Collection: *London Gazette*, July 8, 1710.

[33] Burney Collection: *London Gazette*, May 19, 1716.

[34] Public Record Office PROB 11/539.

[35] Williams (1741, vol. 2, pp. 241–242).

[36] *Oxford Dictionary of National Biography*; biography of Thomas Parker, 1st Earl of Macclesfield, by A. A. Hanham.

[37] De Moivre's incorrect calculation is noted in Hald (1990, p. 531).

[38] Cambridge University Library, Macclesfield Collection, MS Add. 8597/8/9, pp. 79 and 83.

[39] De Moivre (1725, pp. 57–74).

[40] Cambridge University Library, Macclesfield Collection, MS Add. 9597/8/5.

[41] Harrison (1978, p. 193).

[42] Osborne (1742, p. 73).

[43] Cambridge University Library, Macclesfield Collection, MS Add. 9597/8/9, pp. 50–70.

[44] St. John's College Cambridge Library, TaylorB/C3.

[45] Quarrie (2006).

[46] British Library Egerton 922.

[47] Hayes (1727).

[48] Hayes (1718, 1719, 1722, 1724, and 1726). *Young Merchant's Assistant*, written in 1718, contains material on the calculation of customs duties, subsidies, and discounts. The *Negociator's Magazine*, first published in 1719 with a second edition in 1724, contains detailed descriptions of how to calculate exchange rates between various cities and countries as well as the rules for drawing up and executing bills of exchange. The 1722 *Rules for the Port of London* provides all the information necessary to import and export goods including any fees and import or export duties, as well as the names of available wharves. Finally, the *Money'd Man's Guide* from 1726 contains material relevant to annuity calculations. There are tables for annuities for fixed periods of time. The present values of leases for fixed lengths are equated to annuities. Following tradition, annuities based on three lives are equated to a 21-year fixed term annuity. On page 85, there is mention of Halley's and De Moivre's work, but they are not used. The *Young Merchant's Assistant* and *Rules for the Port of London* were both meant to be textbooks for Hayes's students. The fact that the *Negociator's Magazine* went through two editions five years apart says that at least this book was financially successful.

[49] Richards (1730).

[50] The contents of Richards's book have been described in detail by Lewin (2003).

[51] De Moivre (1731), Morris (1735), and Lee (1737).

[52] The sale catalogs for the Macclesfield library sold at Sotheby's auction house can be search at http://www.sothebys.com/app/search/quickSearch/k2/Search.jsp?coll=liveClosed.

[53] Lennon (2006, p. 84).

[54] Hoppit (2003, p. 94).

[55] The additional material appears at the end of the book in an appendix, De Moivre (1731, pp. 111–122).

[56] Berkshire Record Office D/EScv/M/F3–5.

[57] The £700 price for the £100 per annum life annuity is based on the tradition that one life is worth "seven years purchase."

[58] Simpson (1742, p. 39). In the letter, Stevens states he is 47 years of age and refers his correspondent to page 39 of the book he has enclosed with his letter. Page 39 of Simpson is a table of life annuity values. For a person aged 47, the value of an annuity of 1 at 4% interest is 10.5 years purchase. This yields the 1000 guineas for a £100 per annum annuity.

[59] Fielding (1749, p. 97).

[60] Cambridge University Library, Macclesfield Collection, MS Add. 9597/8/9, pp. 23–24 and 75.

[61] Burney Collection: *The London Journal*, August 2, 1729.

[62] Le Blanc (1747, vol. 2, p. 307) mentions De Moivre carrying out calculations on games of chance for gamblers. It is the only contemporary reference to De Moivre offering gambling advice.

[63] Staatsbibliothek zu Berlin, Sig Darmstaedter H 1695: Moivre, Abraham de.

[64] Schneider (2005) dates the manuscript to 1695. However, in correspondence with Iris Lorenz, a librarian at the Staatsbibliothek, 1695 is a catalog number and not a date. Since the manuscript refers to "reversions," it more likely dates from after the publication of *Annuities upon Lives*. Other recorded activities of De Moivre relating to property valuations are also post-1725.

[65] Burney Collection: *London Daily Post and General Advertiser*, July 11, 1739.

[66] There are other examples of promises of confidentiality in these types of business dealings. An advertisement placed in the *London Evening Post*, October 13, 1748 (Burney Collection), by a Mr. B. Scott offers "dispatch and secrecy" to his clients. Scott was a broker or consultant in the buying and selling of estates, annuities on lives, and reversions.

[67] Burney Collection: *London Daily Post and General Advertiser*, March 13, 1741.

[68] British Library, Add. 25103.

[69] *Oxford Dictionary of National Biography*; biography of Joseph Goupy by Sheila O'Connell.

[70] Davis (2010, p. 197).

[71] Vezey (1773, vol. II, pp. 422–423).

[72] The case involved a poor dragoon who held the reversion on an estate. The estate was held by a tenant for life and would go to the tenant's children on his death. If the tenant died without heirs, then the estate would go to the dragoon. The dragoon sold his reversion to another person and the tenant died unmarried one month later.

Chapter 12

[1] De Moivre (1730).

[2] *The Present State of the Republic of Letters*, May 1730, p. 316.

[3] Bellhouse et al. (2009).

[4] A list of members is given in Howells (1982).

[5] Baker (1756, p. 109).

[6] Pearson (2005, p. 19).

[7] Maclaurin (1982, pp. 424–425).

[8] Royal Society Journal Books.

[9] Hall (1980, pp. 2–3).

[10] Walpole (1948, vol. 13, p. 6).

[11] Hazen (1969).

[12] Universitätsbibliothek Basel Bernoulli papers. L I a 22,1, Nr. 56.

[13] Simpson (1740, pp. 7–11).

[14] Dr. Williams's Library, manuscript lecture notes on mathematics L.237 and L.238.

[15] De Moivre (1718, pp. 135–144).

[16] De Moivre (1730, p. 99).

[17] *Oxford Dictionary of National Biography*; biography of Alexander Cuming by Gordon Goodwin and Philip Carter.

[18] Centre for Kentish Studies U1590 C20/4-5. See Bellhouse (2007b) for an analysis.

[19] There are several excellent treatments of De Moivre's derivation of the approximation to the binomial. See, for example, Schneider (1968), Hald (1990), and Stigler (1986).

[20] See Schneider (1968) and Hald (1990) for details on the series expansions and approximations. Stigler (1986) also has an excellent treatment of De Moivre's approximation.

[21] Tweddle (2003, p. 285).

[22] Tweddle (2003, pp. 2–3).

[23] De Moivre (1730, pp. 169–170).

[24] The number is given in De Moivre (1730, p. 170) and Tweddle (2003, p. 135).

[25] De Moivre (1730, supplementum).

[26] Tweddle (2003, pp. 134 and 149).

[27] See Tweddle (2003, pp. 271–272) for a discussion.

[28] De Moivre (1733).

[29] Tweedie (1922, pp. 178–191).

[30] Centre for Kentish Studies U1590/C21.

[31] Sandifer (2007, p. 157).

[32] Daw and Pearson (1972), Schneider (1968).

[33] St. John's College Library, TaylorB/E7.

[34] Royal Society Journal Book, vol. 14, June 14, 1733.

[35] Daniel De Moivre's voyage to Veracruz and his business relationships can be pieced together from documents in the Public Record Office C 104/266, Bundle 38.

[36] McLachlan (1974, pp. 22–24).

[37] Public Record Office C104/266, Bundle 38.

[38] Public Record Office PROB 11/666.

[39] Burney Collection: *London Gazette*, May 13, 1738.

[40] Public Record Office PROB 11/661.

[41] Huguenot Society (1935, p. 35).

[42] *Bibliothéque raisonée des ouvrages des savans de l'Europe*, 1736, pp. 218–219.

[43] For *Miscellanea Analytica*, see *The Present State of the Republick of Letters*, [1730, vol. 5], p. 316. For the second edition of *Doctrine of Chances*, see *London Daily Post and General Advertiser*, July 11 and 18, 1738 (Burney Collection).

[44] There are two Henry Woodfalls, junior and senior. See Plomer et al. (1968, pp. 269–270). They both worked in Paternoster Row and the elder Henry died in about 1747. It is likely that Henry junior was De Moivre's printer for *Doctrine of Chances*. He was the printer for the *London Daily Post and General Advertiser* in which the book was advertised. He and his brother George printed the 1743 edition of *Annuities on Lives*.

[45] Simpson (1742, pp. 4–5).

[46] Burney Collection: *London Daily Post and General Advertiser*, July 11 and18, 1738.

[47] In their analysis of the *Miscellanea Analytica* subscription list, Bellhouse et al. (2009) found subscribers who had died two years before the book was published. According to Montmort's correspondence with Brook Taylor, De Moivre was planning publication of the first edition of *Doctrine of Chances* as early as 1715.

[48] Galiffe (1877, p. 6).

[49] De Moivre (1738, p. 248).

[50] De Moivre (1738, p. 243).

[51] Todhunter (1865, p. 185).

[52] Centre for Kentish Studies U1590 C20/23. See Bellhouse (2007b) for a discussion.

[53] Hartley (1749, pp. 338–339).

[54] Bayes (1763a).

[55] Stigler (1983).

[56] See Bellhouse (2007b).

Chapter 13

1 Royal Society Archives EC/1742/15 and EC/1745/15. In the *Biographia Britannica*, Kippis (1784, p. 218) mentions Simpson as one of Canton's friends. In his letter to John Canton accompanying Bayes's famous paper on probability (Bayes, 1763a), Richard Price refers to Bayes as "our deceased friend."
2 Royal Society Archives, Canton Papers, letter from Bayes to Canton.
3 I have argued in Bellhouse (2004) that what is often used as a portrait of Bayes is imaginative speculation rather than a true portrait. The only picture of Simpson is also doubtful. Both pictures of Bayes and Simpson appear in the same publication, O'Donnell (1936).
4 Janssens and Schillings (2006, p. 104) and Janssens-Knorsch (1990, p. 34).
5 Translation of Lalande (1767, p. 202).
6 Stigler (1986, pp. 89–90) and Hald (1990, pp. 511–512).
7 Burney Collection: *Daily Advertiser*, November 6, 1731.
8 Burney Collection: *London Evening Post*, September 17, 1737.
9 Burney Collection: *London Daily Post and General Advertiser*, March 24, 1738.
10 Todhunter (1865, pp. 204–205) has analyzed Ham's treatment of this problem.
11 Burney Collection: *London Daily Post and General Advertiser*, July 11 and 18, 1738.
12 Burney Collection: *London Daily Post and General Advertiser*, July 24, 1738.
13 I have checked Rocque (1756) as well as keyword searches in Eighteenth Century Collections Online and the Burney Collection of newspapers in the British Library.
14 The first advertisement for the book that I can find is in *London Evening Post*, January 10, 1740. The price is given in a 1741 advertisement in *Daily Gazetteer*, October 5, 1741 (Burney Collection).
15 Simpson (1740, pp. 33–37).
16 Simpson (1740, pp. 59–62).
17 Plomer et al. (1968, pp. 47–48).
18 Simpson (1742). The annuity values are on pages 38–39 and the life table is on pages 4–5.
19 De Moivre (1743, p. xii).
20 Simpson (1742, pp. 49–50).
21 Burney Collection: *London Evening Post*, May 26, 1743.
22 Simpson (1743).
23 Wilkinson (1853).
24 Burney Collection: *Daily Courant*, February 24, 1725.
25 Burney Collection: *Norwich Gazette*, October 24, 1741.
26 Plomer et al. (1968, pp. 171–173).
27 Burney Collection: *General Advertiser*, December 10, 1750.
28 Burney Collection: *London Evening Post*, November 28, 1752.
29 Burney Collection: *Daily Post*, November 22, 1721.
30 Clarke (1929, pp. 39–40).
31 Lalande (1767, p. 202).
32 Holland (1962).
33 Barnard (1958).
34 See, for example, Dale (2003) and Bellhouse (2004).
35 Climenson (1906, p. 18).
36 Newman (1969, p. 105).
37 Centre for Kentish Studies U1590/A98.
38 Weld (1848, p. 460).
39 Centre for Kentish Studies U1590/C21.
40 Maclaurin (1742, pp. 624–627).
41 Bellhouse (2002).
42 De Moivre's three letters to Stanhope are in the Centre for Kentish Studies U1590/C21.
43 Centre for Kentish Studies U1590/A98.
44 Centre for Kentish Studies U1590/C20/23.

[45] Centre for Kentish Studies U1590/C21.
[46] Bayes (1763b).
[47] Royal Society Journal Book, vol. 17.
[48] Royal Society Journal Book, vol. 18.

Chapter 14

[1] Venn and Venn (1922, part I, vol. II, p. 30).
[2] Nichols (1858, vol. 8, pp. 573–574).
[3] Ball (1889, pp. 100–101).
[4] Venn and Venn (1922, part I, vol. I, p. 371).
[5] Burney Collection: *London Evening Post*, May 24, 1739.
[6] Jordan (1735, pp. 147 and 174), Janssens and Schillings (2006, p. 104), and Janssens-Knorsch (1990, p. 34).
[7] *Oxford Dictionary of National Biography*; biography of William Cole by John D. Pickles.
[8] Saunderson (1740, p. vi).
[9] Saunderson (1740, pp. 263–267).
[10] Landesbibliothek Oldenburg, Cim.1, 184.
[11] The answer is $x = 4$ and $y = 1$.
[12] Saunderson (1740, p. 743).
[13] Royal Society Journal Book, vol. 17, January 8, 1740.
[14] Halley (1731).
[15] Cambridge University Library, RGO 14/5, January 1741/2.
[16] Squire (1742).
[17] Royal Society Journal Book, vol. 18, June 7, 1744, and De Moivre (1744).
[18] Pepusch (1746).
[19] Modern musicologists of ancient Greek music say that the Pythagorean system was based on the numbers 1, 2, 3, and 4 instead. See, for example, Ferreira (2002, pp. 2–4).
[20] Burney (1789, vol. IV, p. 638).
[21] Royal Society EC/1745/09.
[22] These are George Lewis Scott (EC/1737/04), John Peter Bernard (EC/1737/09), Philip Naudé (EC/1737/17), Hermann Bernard (EC/1738/11), John Peter Stehelin (EC/1739/10), Peter Davall (EC/1740/07), Henry Steward Stevens (EC/1740/08), Francis Philip Duval (EC/1741/03), Roger Paman (EC/1743//02), Jean Masson (ED/1743/09), Peter Wyche (EC/1745/03), John Christopher Pepusch (EC/1745/09), Edward Montagu (EC/1745/18), Daniel Peter Layard (EC/1746/15), and Daniel Ravaud (EC1747/05).
[23] Royal Society EC/1753/07 and Heilbron (1976).
[24] Le Blanc (1747, vol. 1, pp. 168–169).
[25] Maty (1755) with translations in Bellhouse and Genest (2007, p. 127), Formey (1748), and Royal Society EC/1737/17.
[26] Stephen and Lee (1922, vol. 1, p. 88).
[27] Tardy (1800, p. 250).
[28] Bibliothèque nationale de France, Procès verbaux de l'Académie royale des sciences, tome 73, 1754, pp. 425 and 428.
[29] Maty (1755).
[30] Burney Collection: *Whitehall Evening News or London Intelligencer*, August 24, 1754, and *Public Advertiser*, August 26, 1754.
[31] Burney Collection: *Public Advertiser*, July 31, 1755.
[32] Centre for Kentish Studies U1590/C21.
[33] Maty (1755). See also Bellhouse and Genest (2007, p. 129).
[34] Burney Collection: *General Advertiser*, May 15, 1751.
[35] Maty (1755) as translated in Bellhouse and Genest (2007, p. 129).

[36] Parish register, St. Martin-in-the-Fields Church, 1754.
[37] Public Record Office PROB/11/811.
[38] Heilbron (1976).
[39] Public Record Office PROB/11/950.
[40] Burney Collection: *London Evening Post,* December 10, 1754.
[41] De Moivre also frequented Pons Coffeehouse. See Climenson (1906, p. 286).
[42] Burney Collection: *London Evening Post*, September 20, 1755.
[43] See the online *Oxford Dictionary of National Biography* under the entry for James Dodson by G. J. Gray and Anitia McConnell.
[44] Le Blanc (1747, vol. II, p. 309).
[45] Smiles (1868, pp. 237–238).
[46] Maclaurin (1982, pp. 412–419).
[47] See Grattan-Guiness (1969) for a discussion.
[48] Centre for Kentish Studies U1590/C20/5 and U1590/C20/8.
[49] Burney Collection: *London Evening Post*, January 24, 1756.
[50] Burney Collection: *Public Advertiser*, July 28, 1760.
[51] Gaskell (1957).
[52] Maclaurin (1748).
[53] Wood (2003, p. 102).
[54] De Moivre (1756, p. xi).
[55] Centre for Kentish Studies U1590/C14/2.
[56] Bellhouse (2007b).
[57] De Moivre (1756, p. 332).
[58] Hald (1990, p. 391).
[59] Bayes's death notice appears in *Whitehall Evening Post or London Intelligencer,* April 14, 1761; Folkes's death notice is in *Public Advertiser,* July 1, 1754; and Macclesfield's death notice is in *London Chronicle*, March 20, 1764 (Burney Collection).

Bibliography

Manuscript Sources

Berkshire Record Office
> Papers of Henry Stuart Stevens
> D/ESv/M/F3–5

La Bibliothèque de la Société de l'Histoire du Protestantisme français
> Ms 171. Copie, exécutée en 1783, d'un recueil de Jacob Varnier sur les familles protestante de Vitry-le-François, au moment de la Révocation (material on the Moivre family, p. 171).

Bibliothèque nationale de France
> Procès verbaux de l'Académie royale des sciences, tome 73, 1754, pp. 425 and 428.

British Library Manuscripts
> Add. 4281–4289. Correspondence and papers of Pierre Des Maizeaux.
> Add. 25103. Album of Cox Macro, D.D. of Norton in Suffolk.
> Egerton 922. Copies of various statistical and political papers, relating to Great Britain and France, said to have belonged to Sir Robert Walpole.

Cambridge University Library
> Macclesfield Collection
>> MS Add. 9597/4/9 (Combinations—rough copy).
>> MS Add. 9597/4/10 (Combinations—fair copy).
>> MS Add. 9597/4/14 (Reasonings Concerning Chance).
>> MS Add. 9597/4/19 (Sir Isaac Newton's Questions).
>> MS Add. 9597/8/5 (De Moivre's solution of a Question concerning the value of lives).
>> MS Add. 9597/8/9 (Annuities Upon Lives).
>> MS Add. 9597/8/10 (Simple Interest).
>> MS Add. 9597/8/11 (Of Annuities in Reversion).
>> MS Add. 9597/8/12 (Simple Interest, Of Purchasing freehold Estates by the years purchase and Compound Interest).

Portsmouth Collection
   MS Add. 4000 (College Notebook).
   MS Add. 4005.28: 102–103 (A Problem in Chances).
   Royal Greenwich Observatory Archives
   RGO 14/5 (Papers of the Board of Longitude: confirmed minutes).
Carnegie Mellon University
   Posner Family Collection
   523.6 H18A *Astronomiae Cometicae Synopsis. Autore Edmundo Halleio.*
Centre for Kentish Studies
   U1590/A98: Account of my Expenses beginning November the 28th. 1735.
   U1590/C14/2: Letters from Patrick Murdoch to Lord Stanhope.
   U1590/C20/4: Analytical Problems.
   U1590/C20/5: Analytical & Other Theorems.
   U1590/C20/8: Annotations on several analytical authors.
   U1590/C20/23: Annotations on Sundry Works on Chances.
   U1590/C21. Papers by several eminent mathematicians addressed to or collected by
   Lord Stanhope (contains three letters from De Moivre to Stanhope and several papers
   by Thomas Bayes).
Columbia University Library
   Special Collections
   David Eugene Smith Collection: letter from Abraham de Moivre to Edward Montague.
Dr. Williams's Library
   Manuscript Collection
   Congregational Library I.e.28 (Notes on Mathematics, Natural Philosophy, Logic,
   Theology, etc.).
   MS NCL/L.237 (Lecture notes on mathematics).
   MS NCL/L.238 (Lecture notes on mathematics).
Edinburgh University Library Special Collections Division
   Papers of David Gregory
   GB 0237 David Gregory Dk.1.2.2 Folio B [18] (A treatise of chance written by Dr
   Arbuthnot anno 1694).
Huguenot Society Library, University College London
   Manuscript collection
   Conversions et reconnoissances faites à l'Église de la Savoye 1684–1702.
London Metropolitan Archives
   MS 12 819/12 (Christ's Hospital Treasurer's Account Books 1695–1717).
   St. Giles-in-the-Fields burial register.
Public Record Office
   C 104/266 Bundle 38: Papers of Daniel de Moivre relating to trade, mainly in precious
   stones and jewellery, in London and Vera Cruz, Mexico.
   PROB 11/539: will of Catherine Thompson.
   PROB 11/588: will of Peter de Magneville.
   PROB 11/653: will of Thomas Woodcock.
   PROB 11/661: will of Daniel De Moivre (senior).
   PROB 11/666: will of Daniel De Moivre (junior).
   PROB 11/811: will of Abraham De Moivre.
   PROB/11/950: will of John Gray.

Royal Society Archives
  Classified Papers
    Cl.P/1/4: "Questions of Chance" by Mr Robartes.
  Canton Papers
    Correspondence, vol. 2, folio 32 (letter from Thomas Bayes to John Canton).
  Early Letters
    EL/D2/6: Richard Dobbs, Trinity College, Dublin to James Jurin.
    EL/D2/7: Letter from Richard Dobbs, Castle Dobbs, County Antrim.
    EL/H3/53: Report of the committee for inspecting the books and papers of the Royal Society, dated between 1669–1677, in relation to the dispute between Leibniz and Keil.
  Election Certificates
    EC/1737/04 (George Lewis Scott).
    EC/1737/09 (John Peter Bernard).
    EC/1737/17 (Philip Naudé).
    EC/1738/11 (Hermann Bernard).
    EC/1739/10 (John Peter Stehelin).
    EC/1740/07 (Peter Davall).
    EC/1740/08 (Henry Stewart Stevens).
    EC/1741/03 (Francis Philip Duval).
    EC/1742/15 (Thomas Bayes).
    EC/1743/02 (Roger Paman).
    EC/1743/09 (Jean Masson).
    EC/1745/03 (Peter Wyche).
    EC/1745/09 (John Christopher Pepusch).
    EC/1745/15 (Thomas Simpson).
    EC/1745/18 (Edward Montagu).
    EC/1746/15 (Daniel Peter Layard).
    EC/1747/05 (David Ravaud).
    EC/1753/07 (Robert Symmer).
  Journal Books of Scientific Meetings: Volumes 9, 10, 11
Staatsbibliothek zu Berlin—Preußischer Kulturbesitz Handschriftenabteilung
  Sig. Darmstaedter H 1695: Moivre, Abraham de.
St. John's College Library, Cambridge
  TaylorB/C3 (Manuscript notes relating to annuities).
  TaylorB/E4 (Autograph letters from Pierre Rémond de Montmort to Taylor).
  Taylor B/E7 (Correspondence between Taylor and Abraham de Moivre).
Universitätsbibliothek Basel
  Bernoulli Papers
    L I a 22,1, Nr. 56. (Gabriel Cramer to Nicolaus I Bernoulli 1731.06.12).
    L I a 22,2, Nr. 180a. (Abraham De Moivre to Nicolaus I Bernoulli 1714.03.03).
    L I a 654, Nr. 9. (Johann I Bernoulli to William Burnet 1712.03.05).
    L I a 654, Nr. 10. (Johann I Bernoulli to William Burnet 1712.08.24).
    L I a 654, Nr. 11. (Johann I Bernoulli to William Burnet 1712.12.19).
    L I a 654, Nr. 13. (Johann I Bernoulli to William Burnet 1714.02.19).
    L I a 654, Nr. 19*. (William Burnet to Johann I Bernoulli 1714.04.08).
    L I a 665, Nr. 7. (Johann I Bernoulli to Pierre Rémond de Montmort 1717.04.08).
    L I a 665, Nr. 11. (Johann I Bernoulli to Pierre Rémond de Montmort 1718.05.21).

L I a 665, Nr. 12. (Johann I Bernoulli to Pierre Rémond de Montmort 1718.09.29).
L I a 665, Nr. 13*. (Pierre Rémond de Montmort to Johann I Bernoulli 1716.12.28).
L I a 665, Nr. 15. (Johann I Bernoulli to Pierre Rémond de Montmort 1719.07.13).
L I a 665, Nr. 22*. (Pierre Rémond de Montmort to Johann I Bernoulli 1719.06.12).
L I a 673, Nr. 1, fol. 5–6. (Johann I Bernoulli to John Arnold 1714.03.08).
L I a 673, Nr. 1, fol. 7–8. (Johann I Bernoulli to John Arnold 1714.04.05).
L I a 673, Nr. 1, fol. 9–10. (Johann I Bernoulli to John Arnold 1714.05.09).
L I a 673, Nr. 1, fol. 13–14. (Johann I Bernoulli to John Arnold 1715.11.09).
Landesbibliothek Oldenburg
Cim.1, 184 (Leçons sur Algêbre par Mons: Abr: de Moivre à Londres commence le 7me du Mai 1742 fine dans le mois d'Avril 1743).
University of Chicago
The Joseph Halle Schaffner Collection box 1, folder 51 (Memorandum relating to Sir Isaac Newton given me by Mr. Abraham Demoivre in Novr. 1727).
Wellcome Library for the History & Understanding of Medicine
Western MS. 6146/56 (Letter to Richard Dobbs from James Jurin).
Westminster Council Archives
Parish Register, St. Martin-in-the-Fields Church, 1754.
Poor Law Rate Books, St. Martin-in-the-Fields Parish, 1750–1754.

Periodicals

Bibliothéque raisonée des ouvrages des savans de l'Europe. Amsterdam, Wetsteins & Smith, 1728–1753.
Journal de Trévoux ou memoires pour server a l'histoire des sciences et des arts, Trévoux, 1701–1767.
Journal literaire. La Haye, T. Johnson, 1713–1737.
Present State of the Republick of Letters, London, Innys, 1728–1736.

Online Sources

British History Online
http://www.british-history.ac.uk/
Mapping the Republic of Letters
https://republicofletters.stanford.edu
Oxford University Press. Oxford Dictionary of National Biography.
http://www.oxforddnb.com/
Oxford University Press. Oxford English Dictionary Online.
http://dictionary.oed.com/
Gale Digital Collections: 17th–18th Century Burney Collection Newspapers
http://gdc.gale.com/products/17th-and-18th-century-burney-collection-newspapers/

Printed Sources

Aguilón, F. de (1613). *Opticorum libri sex, Philosophis juxta ac mathematicis utiles.* Antwerp: Plantin.

Agnew, D. C. A. (1874). *Protestant Exiles from France in the Reign of Louis XIV*. London: Reeves and Turner.

Anonymous (1651). *The Royall and Delightfull Game of Picquet, written in French and Now Rendred into English out of the Last French Edition*. London: J. Martin and J. Ridley.

Anonymous (1709). Review of *Essay d'analyse sur les jeux de hazard*, by Pierre Rémond de Montmort, *Supplément du Journal des sçavans*: 433–471.

Anonymous (1731). *A View of the Town: or, Memoirs of London*. London: A. Moore.

Anonymous (1949). The Society's portraits. *Notes and Records of the Royal Society* 7: 97–107.

Arbuthnot, J. (1692). *Of the Laws of Chance, or, a Method of Calculating the Hazards of Game*. London: Motte.

Arbuthnott, J. (1710). An argument for Divine Providence, taken from the constant regularity observ'd in the births of both sexes. *Philosophical Transactions* 27: 186–190.

Archibald, R. C. (1929). A letter of De Moivre and a theorem of Halley. *The Mathematical Gazette* 14: 574–575.

Archibald, R. C. (1945). First published mortality table. *Mathematical Tables and Other Aids to Computation* 1: 402–403.

Arnstein, W. L. (1993). *The Past Speaks: Sources and Problems in British History*. vol. II. second edition. Lexington, MA: Heath.

Baker, S. (1756). *A Catalogue of the Entire and Valuable Library of Martin Folkes*. London, Samuel Baker.

Ball, W. W. R. (1889). *A History of Mathematics at Cambridge*. Cambridge: Cambridge University Press.

Ball, W. W. R. (1912). *A Short Account of the History of Mathematics*. London: Macmillan.

Barnard, G. A. (1958). Thomas Bayes—a biographical note. *Biometrika* 45: 293–295.

Barnard, H. C. (1922). *French Tradition in Education: Ramus to Mme. Necker de Saussure*. Cambridge: Cambridge University Press.

Barrow, I. (1685). *Issaci Barrow, Lectiones Mathematicae XXIII, in quibus Principia Matheseôs Generalia Exponuntur Habitae Cantabrigiae A.D. 1664, 1665, 1666*. London: John Playford.

Barrow, I. (1734). *The Usefulness of Mathematical Learning Explained and Demonstrated: Being Mathematical Lectures Read in the Publick Schools at the University of Cambridge*. London: Stephen Austen.

Barthélemy, É. (1861). *Diocèse ancien de Châlons-sur-Marne, histoire et monuments: suivi des cartulaires inédits de la commanderie de la Neuville-au-Temple, des abbayes de Toussaints, de Monstiers et du prieuré de Vinetz*. Paris: A. Aubry.

Bayes, T. (1763a). An essay towards solving a problem in the doctrine of chances. *Philosophical Transactions of the Royal Society* 53: 370–418.

Bayes, T. (1763b). A letter from the late Reverend Mr. Thomas Bayes to John Canton, M.A. & F.R.S. *Philosophical Transactions of the Royal Society* 53: 269–271.

Bellhouse, D. R. (2004). The Reverend Thomas Bayes, FRS: A biography to celebrate the tercentenary of his birth. *Statistical Science* 19: 3–43.

Bellhouse, D. R. (2007a). The Problem of Waldegrave. *Journal Electronique d'Histoire des Probabilités et de la Statistique*, 3 (2) http://www.jehps.net/decembre2007.html.

Bellhouse, D. R. (2007b). Lord Stanhope's papers on the doctrine of chances. *Historia Mathematica* 34: 173–186.

Bellhouse, D. R. (2008). Banishing Fortuna: Montmort and De Moivre. *Journal of the History of Ideas* 69: 559–581.

Bellhouse, D. R. (2011). A new look at Halley's life table. *Journal of the Royal Statistical Society, (A)* 174: 823–832.

Bellhouse, D. R. and Davision, M. (2009). De Moivre's Poisson approximation to the binomial. *International Statistical Review* 77: 451–449.

Bellhouse, D. R. and Genest, C. (2007). Maty's biography of Abraham De Moivre, translated, annotated and augmented. *Statistical Science* 22, no. 1: 109–136.

Bellhouse, D. R., Renouf, E. M., Raut, R., and Bauer, M. A. (2009). De Moivre's knowledge community: An analysis of the subscription list to the *Miscellanea Analytica. Notes and Records of the Royal Society* 63: 137–162.

Berkeley, G. (1734). *The Analyst; or, a Discourse Addressed to an Infidel Mathematician.* London: Tonson.

Bernoulli, Ja. (1713) *Ars Conjectandi.* Basel: Thurnisius.

Bernoulli, Jo. (1710). Extrait d'une lettre de M. Herman à M. Bernoulli datée de Padoüe le 12 juillet 1710. *Histoire de l'académie royale des sciences. Année M.DCC.X. avec les mémoires de mathématique et de physique pour la même année*, pp. 519–533. Paris, 1732.

Bernoulli, Jo. (1716). Epistola pro eminente mathematico, Dn Johanne Bernoullio, contra quendam ex Anglia antagonistam scripta. *Acta Eruditorum*, pp. 296–315.

Bernoulli, Jo. (1992). *Der Briefwechsel von Johann I Bernoulli. Band 3, Der Briefwechsel mit Pierre Varignon Teil II: 1703–1714*, Pierre Costabel, Jeanne Peiffer, Fritz Nagel and Martin Mattmüller (eds.). Basel: Birkhäuser.

Bernoulli, N. (1714). Solutio generalis Problematis XV. Propositi a D. de Moivre, in tractatu de mensura sortis inserto Actis Philosophicis Anglicanis No 329. Pro numero quocunque collusorum. *Philosophical Transactions* 29: 133–144.

Bertrand, G. (1874). *Catalogue des manuscrits français de la bibliothèque de Saint-Pétersbourg.* Paris: Imprimerie Nationale.

Bickley, F. (1911). *The Cavendish Family.* London: Constable.

Bossut, C. (1803). *A General History of Mathematics from Earliest Times to the Middle of the Eighteenth Century.* London: J. Johnson.

Bottazzini, U. (1996). Introduction to the mathematical writings from Daniel Bernoulli's youth. In Die Werke von Daniel Bernoulli, Band 1, pp. 129–194. Basel: Birkhäuser.

Bourchenin, P.-D. (1882). *Étude sur les académies protestantes en France au XVIe et au XVIIe siècle.* Paris: Grassart.

Boyer, A. (1700). *Achilles, or, Iphigenia in Aulis.* London: Bennet.

Brewster, D. (1855). *Memoirs of the Life, Writings and Discoveries of Sir Isaac Newton.* Edinburgh: Constable.

Burney, C. (1789). *A General History of Music from the Earliest Ages to the Present Period.* Volume IV. London.

Burtchaell, G. D. and Sadleir, T. U. (1924). *Alumni Dublinenses: A Register of the Students, Graduates, Professors and Provosts of Trinity College in the University of Dublin.* London: Williams and Norgate.

Centlivre (1705). *The Basset Table, a Comedy.* London: Turner.

Challis, C. E. (1992). *A New History of the Royal Mint.* Cambridge: Cambridge University Press.

Chamberlayne, J. (1710). *Magnæ Britanniæ Notitia: or, the Present state of Great-Britain, With Divers Remarks upon the Ancient State thereof.* Twenty-third edition. London: Goodwin.

Chamberlayne, J. (1716). *Magnæ Britanniæ Notitia: or, the Present state of Great-Britain, With Divers Remarks upon the Ancient State thereof.* Twenty-fourth edition. London: Goodwin.

Chambers, E. (1778–1788). *Cyclopædia: or, an Universal Dictionary of Arts and Sciences*, vol. 3. London: W. Strahan.

Cherry, B. (2001). Edward Hatton's *New View of London*. *Architectural History* 44: 96–105.

Cheyne, G. (1703). *Fluxionum Methodus Inversa: sive Quantitatum Fluentium Leges Generales*. London: J. Matthew.

Cheyne, G. (1705). *Rudimentorum Methodi Fluxionum Inversæ Specimina: quæ responsionem continent ad animadversiones ab. de moivre in librum G. Chæynei, M.D. S.R.S.* London: Benjamin Motte.

Cheyne, G. (1724). *An Essay of Health and Long Life*. London: Strahan.

Childs, J. (1991). *The Nine Years' War and the British Army, 1688–1697: the Operations in the Low Countries*. Manchester: Manchester University Press.

Cibber, C. (1728). *The Provok'd Husband; or a Journey to London*. London: J. Watts.

Clark, G. W. (1999). *Betting on Lives: The Culture of Life insurance in England, 1695–1775*. Manchester: Manchester University Press.

Clarke, F. M. (1929). *Thomas Simpson and His Times*. New York: Waverley Press.

Claude, J. (1708). *A Short Account of the Complaints, and Cruel Persecutions of the Protestants in the Kingdom of France*. London: W. Redmayne.

Climenson, M. J. (1906). *Elizabeth Montagu: the Queen of the Bluestockings, her Correspondence from 1720–1761*. London: John Murray.

Cohen, I. B. (1985). *Revolution in Science*. Cambridge, MA.: Belknap Press of Harvard University Press.

Cook, A. (1993). Halley the Londoner. *Notes and Records of the Royal Society of London* 47: 163–177.

Cook, A. (1998). *Edmond Halley: Charting the Heavens and the Seas*. Oxford: Clarendon Press.

Cooper, W. D. (1862). *Lists of Foreign Protestants, and Aliens, Resident in England 1618–1688. From Returns in the State Paper Office*. London: Camden Society.

Cotton, C. (1674). *The Compleat Gamester*. London: Brome.

Cotton, C. (1709). *The Compleat Gamester*. London: Brome.

Cowan, B. W. (2005). *The Social Life of Coffee: The Emergence of the British Coffeehouse*. New Haven: Yale University Press.

Dale, A. I. (2003). *Most Honourable Remembrance: The Life and Work of Thomas Bayes*. New York: Springer.

Daston, L. (1988). *Classical Probability in the Enlightenment*. Princeton: Princeton University Press.

Daumas, M. (1969). *A History of Technology & Invention: Progress Through the Ages*. Vol. 2 of *The First Stages of Mechanization* (trans. E. B. Hennessy). New York: Crown Publishers.

David, F. N. (1962) *Games, Gods and Gambling: A History of Probability and Statistical Ideas*. London: Charles Griffin.

Davis, G. (2010). *A Companion to Horace*. Chichester: Blackwell.

Daw, R.H. and Pearson, E. S. (1972). Abraham De Moivre's 1733 derivation of the normal curve: A bibliographical note. *Biometrika* 59: 677–680.

Dechales, C.-F. M. (1685). *The elements of Euclid explain'd, in a new, but most easie method together with the use of every proposition through all parts of the mathematicks*. Oxford: Lichfield.

De Moivre, A. (1695). De specimina quaedam illustria doctrinae fluxionum sive exempla quibus methodi istius usus et praestantia in solvendis problematis geometricis elucidatur, ex epistola peritissimi mathematici D. Ab. de Moivre desumpta. *Philosophical Transactions* 19: 55–57.

De Moivre, A. (1697). A method of raising an infinite multinomial to any given power, or extracting any given root of the same. *Philosophical Transactions* 19: 619–625.

De Moivre, A. (1698). A method of extracting the root of an infinite equation. *Philosophical Transactions* 20: 190–193.

De Moivre, A. (1700). The dimensions of solids generated by the conversions of Hippocrates's lunula, and of its parts about several axes, with the surfaces generated by that conversion. *Philosophical Transactions* 22: 624–626.

De Moivre, A. (1702). Methodus quandrandi genera quaedam curvarum, aut ad curvas simpliciores reducendi. *Philosophical Transactions* 23: 1113–1127.

De Moivre, A. (1704). *Animadversiones in G. Cheynaei Tractatum de Fluxionum Methodo Inversa.* London: Edward Midwinter.

De Moivre, A. (1707). Aequationum quarundam potestatis tertiae, quintae, septimae, nonae, & superiorum, ad infinitum usque pergendo, in terminis finitis, ad instar regularum pro cubicus quae vocantur Cardani, resolution analytica. *Philosophical Transactions* 25: 2368–2371.

De Moivre, A. (1711). De mensura sortis seu; de probabilitate eventum in ludis a casu fortuito pendentibus. *Philosophical Transactions* 27: 213–264.

De Moivre, A. (1714). Solutio generalis altera praecedentis, ope combinationum & serierum infinitarum. *Philosophical Transactions* 29: 145–158.

De Moivre, A. (1715). A ready description and quadrature of a curve of the third order, resembling that commonly call'd the foliate. *Philosophical Transactions* 29: 329–331.

De Moivre, A. (1717). Proprietates quaedam simplices sectionum conicarum ex natura focorum deductae; cum theoremate generali de viribus centripetis; quorum ope lex virium centripetarum ad focos sectionum tendentium, velocitates corporum in illis revolventium, & descriptio orbium facillime determinantur. *Philosophical Transactions* 30: 622–628.

De Moivre, A. (1718). *The Doctrine of Chances, or, a Method of Calculating the Probability of Events in Play.* London: W. Pearson.

De Moivre, A. (1719). De maximis & minimis quae in motibus corporum coelestium occurrunt. *Philosophical Transactions* 30: 952–954.

De Moivre, A. (1722a). De fractionibus algebraicis radicalitate immunibus ad fractiones simpliciores reducendis, deque summandis terminis quarumdam serierum aequali intervallo a se distantibus. *Philosophical Transactions* 32 162–178.

De Moivre, A. (1722b). De sectione anguli. *Philosophical Transactions* 32: 228–230.

De Moivre, A. (1725). *Annuities upon Lives, or, The valuation of Annuities upon any Number of Lives, as also, of Reversions to which is added, an Appendix Concerning the Expectations of Life, and Probabilities of Survivorship.* London: W. P.

De Moivre, A. (1730). *Miscellanea Analytica de Seriebus et Quadraturis.* London: J. Tonson and J. Watts.

De Moivre, A. (1731). *Annuities upon Lives, or, The valuation of Annuities upon any Number of Lives, as also, of Reversions to which is added, an Appendix Concerning the Expectations of Life, and Probabilities of Survivorship.* Second edition, corrected. Dublin: Samuel Fuller.

De Moivre, A. (1733). *Approximatio ad Summam Terminorum Binomii* $\overline{a+b}^n$ *in Seriem expansi.* London—n.p.

De Moivre, A. (1738). *The Doctrine of Chances: or, a Method of Calculating the Probability of the Events in Play,* second edition. London: Woodfall.

De Moivre, A. (1743). *Annuities on Lives: Second Edition, Plainer, Fuller, and More Correct than the Former*. London: Woodfall.

De Moivre, A. (1744). A letter from Mr. Abraham De Moivre, F. R. S. to William Jones, Esquire, F. R. S. concerning the easiest method for calculating the value of annuities upon lives, from tables of observations. *Philosophical Transactions* 43: 65–78.

De Moivre, A. (1750). *Annuities on Lives: Third edition, Plainer, Fuller, and More Correct than the Former*. London: Millar.

De Moivre, A. (1752). *Annuities on Lives: with Several Tables, Exhibiting at One View, the Values of Lives, for Different Rates of Interest*. London: Millar.

De Moivre, A. (1756). *The Doctrine of Chances: Or, a Method of Calculating the Probability of Events in Play*, third edition. London: Millar. Reprinted, 1967 by Chelsea Publishing Co., New York.

De Moivre, A. (1984). On the measurement of chance, or, on the probability of events in games depending upon fortuitous chance (Bruce McClintock trans.). *International Statistical Review* 52: 237–262.

De Moivre, A. (2009). *Mélange analytiques. Suivi de remarques sur la méthode inverse inverse des fluxions de Georges Cheyne*. Jean Peyroux (trans.). Paris: Blanchard.

Des Chene, D. (2002). Cartesian Science: Régis and Rohault. In: *A Companion to Early Modern Philosophy* (ed. S. M. Nadler). Malden, MA: Blackwell Publishing, pp. 183–196.

Deslande, A. F. (1713). *Poetae Rustiantis Literatum Otium*. London: Lintott.

Deslande, A. F. (1737). *Histoire critique de la philosophie*. Amsterdam: François Changuion.

Des Maizeaux, P. (1720). *Recueil de diverses pièces, sur la philosophie, la religion naturelle, l'histoire, les mathématiques, &c. Par Mrs. Leibniz, Clarke, Newton, & autres autheurs célèbres*, vol. II. Amsterdam: Sauzet.

De Vries, J. and Van Der Woude, A. (1997). *The First Modern Economy: Success, Failure, and Perseverance of the Dutch Economy, 1500–1815*. Cambridge: Cambridge University Press.

Dobbs, B. J. T. (1975). *The Foundations of Newton's Alchemy, or "The Hunting of the Greene Lyon."* Cambridge: Cambridge University Press.

Dodson, J. (1742). *The Anti-Logarithmic Canon*. London: Dodson and Wilcox.

Douen, O. (1894). *Le Révocation de l'Édit de Nantes à Paris d'après des document inédits*, vol. 3. Paris: Fischbacher.

Edleston, J. (1969). *Correspondence of Sir Isaac Newton and Professor Cotes*. London: F. Cass.

Edwards, A. W. F. (1987). *Pascal's Arithmetical Triangle*. London: Griffin.

Edwards, C. H. (1979). *The Historical Development of the Calculus*. New York: Springer.

Englesman, S. B. (1991). *Families of Curves and the Origins of Partial Differentiation*. Amsterdam: Elsevier.

Euclid (1570). *The Elements of Geometrie of the most Auncient Philosopher Euclide of Megara*. H. Billingsley (trans.). London: Daye.

Evans, W. D. (1817). *A Collection of Statutes Connected with the General Administration of the Law*, vol. I. London: Butterworth.

Ewles-Bergeron, P. (1997). Edward Mangin's list of Huguenot names, 1841. *Proceedings of the Huguenot Society* 26, no. 5: 611–634.

Fallon, J. P. (1972). *Marks of London Goldsmiths and Silversmiths: Georgian Period (c 1697–1837)*. New York: ARCO Publishing.

Farquhar, G. (1701). *Sir Henry Wildair: Being a Sequel of the Trip to the Jubilee*. London: James Knapton.

Fatio de Duillier, N. (1699). *Lineae Brevissimi Descensus Investigatio Geometrica Duplex*. London: Everingham.

Feigenbaum, L. (1985). Brook Taylor and the method of increments, *Archive for History of Exact Sciences* 34: 1–140.

Ferraro, G. (2008). *The Rise and Development of the Theory of Series up to the Early 1820s*. New York: Springer.

Ferreira, M. P. (2002). Proportions in ancient and medieval music. In *Mathematics and Music*, G. Assayag, H. G. Feichtinger, and J. H. Rodrigues (eds.), pp. 1–25. New York: Springer.

Fielding, H. (1749). *The History of Tom Jones, a Founding*. London: A. Millar.

Flamsteed, J. (1995–2002). *The Correspondence of John Flamsteed, the First Astronomer Royal*, 3 volumes. E. G. Forbes, L. Murdin, and F. Willmoth (eds.). Bristol and Philadelphia: Institute of Physics Publishing.

Folkerts, M. (2004). Eine algebravorlesung von Abraham de Moivre. In *Form, Zahl, Ordnung*: *Studien zur Wissenschafts- und Technikgeschicte*, pp. 269–275. R. Seising, M. Folkerts, U. Hashagen (eds.). Stuttgart: Franz Steiner Verlag.

Fontenelle, B. (1719). Éloge de M. de Montmort. *Histoire de l'Académie royale des sciences. Année 1719* (Paris, 1721), pp. 83–93.

Formey, S. (1748). Éloge de Monsieur Naudé. *Histoire de l'Académie royale des sciences et de belles lettres de Berlin, 1746*. Berlin: Ambroise Haude.

Foster, J. (1968). *Alumni Oxonienses, the Members of the University of Oxford*. Lichtenstein: Kraus Reprint.

Francis, J. (2008). *Philosophy of Mathematics*. New Delhi: Global Vision Publishing House.

Galiffe, J.-B.-G. (1877). *D'un siècle a l'autre: correspondences inédites entre gens connus et inconnus du XVIIIe et du XIXe siècle*. Geneva: Jules Sandoz.

Galloway, T. (1839). *A Treatise on Probability: Forming the Article under that Head in the Seventh Edition of the Encyclcopædia Britannica*. Edinburgh: Adam and Charles Black.

Garnier, A. E. (1900). *Chronicles of the Garniers of Hampshire During Four Centuries, 1530–1900*. Norwich and London: Jarrold and Sons.

Gaskell, P. (1957). Notes on eighteenth-century British paper. *Library* S5–XII: 34–42.

Gater, G. H. and Hiorns, F. R. (1940) *Survey of London: Volume 20 Trafalgar Square and Neighbourhood (The Parish of St Martin-in-the-Fields, Part III)*, G. H. Gater and W. H. Godfrey (eds.). London: London County Council.

Goldbach, C. (1720). Specimen method at summas serierum. *Acta Eruditorum* 39: 27–31.

Gowing, R. (1983). *Roger Cotes—Natural Philosopher*. Cambridge: Cambridge University Press.

Grafton, A., 2009. A sketch map of a lost continent: The Republic of Letters. *Republics of Letters*: *A Journal for the Study of Knowledge, Politics, and the Arts* 1. http://rofl.stanford.edu/node/34.

Grattan-Guiness, I. (1969). Berkeley's criticism of the calculus as a study in the theory of limits. *Janus* 56: 215–227.

Gravelle, T. L. and Miller, G. (1983). *A Catalogue of Foreign Watermarks Found on Paper Used in America*. New York and London: Garland Publishing.

Gregory, D. (1937). *David Gregory, Isaac Newton and their Circle*: *Extracts from David Gregory's Memoranda 1677–1708*. W. G. Hiscock (ed.) Oxford: Oxford University Press.

Guerrini, A. (1986). The Tory Newtonians: Gregory, Pitcairne, and their circle. *The Journal of British Studies* 25: 288–311.

Guerrini, A. (2000). *Obesity and Depression in the Enlightenment: The Life and Times of George Cheyne.* Norman, OK: University of Oklahoma Press.

Guicciardini, N. (1989). *The Development of Newtonian Calculus in Britain 1700–1800.* Cambridge: Cambridge University Press.

Guicciardini, N. (1995). Johann Bernoulli, John Keill and the inverse problem of central forces. *Annals of Science* 52: 537–575.

Guicciardini, N. (1999). *Reading the Principia: The Debate on Newton's Mathematical Methods for Natural Philosophy from 1687 to 1736.* Cambridge: Cambridge University Press.

Guicciardini, N. (2004). Isaac Newton and the publication of his mathematical manuscripts. *Studies in the History and Philosophy of Science* 35: 455–470.

Gwynn, R. D. (1985). *Huguenot Heritage: The history and contribution of the Huguenots in Britain.* London: Routledge and Kegan Paul.

Haag, M. & Haag, E. (1846–1859). *La France protestante.* Paris: J. Cherbuliez. Reprinted in 1966 by Slatine Reprints, Geneva, Switzerland.

Hald, A. (1990). *A History of Probability and Statistics and Their Applications before 1750.* New York: Wiley.

Hall, A. R. (1980). *Philosophers at War: The Quarrel between Newton and Leibniz.* Cambridge: Cambridge University Press.

Halley, E. (1693). An estimate of the degrees of mortality of mankind, drawn from the curious tables of births and funerals in the City of Breslaw; with an attempt to ascertain the price of annuities upon lives. *Philosophical Transactions* 17: 596–610.

Halley, E. (1696). An easie demonstration of the analogy of the logarithmick tangents to the meridian line or sum of the secants: with various methods for computing the same to the utmost exact ness. *Philosophical Transactions* 19: 202–214.

Halley, E. (1700). De iride, sive de arcu caelesti, differtatio geometrica, qua methodo directa iridis ntriusq; diameter, data ratione refractionis, obtinetur: cum solutione inversi problematis, sive inventione rationis istius ex data arcus diametro. *Philosophical Transactions* 22: 714–725.

Halley, E. (1706). *Miscellanea Curiosa: Being a Collection of Some of the Principal Phaenomena in Nature, Accounted for by the Greatest Philosophers of this Age. Together with Several Discourses Read Before the Royal Society, for the Advancement of Physical and Mathematical Knowledge,* vol. 2. London: J. B.

Halley, E. (1715). Observations of the late total eclipse of the sun on the 22d of April last past, made before the Royal Society at their house in Crane-Court in Fleet-Street, London. *Philosophical Transactions* 29: 245–262.

Halley, E. (1731). A proposal of a method for finding the longitude at sea within a degree, or twenty Leagues. *Philosophical Transactions* 37: 185–195.

Harris, J. (1710). *Lexicon technicum: or, an Universal English Dictionary of Arts and Sciences: Explaining Not only the terms of Art, but the Arts themselves,* vol. II. London: Brown et al.

Harrison, J. (1978). *The Library of Isaac Newton.* Cambridge: Cambridge University Press.

Hartley, D. (1749). *Observations on Man, his Frame, his Duty, and his Expectations.* London: Richardson.

Harvey, S. and Grist, E. (2006). The Rainbow Coffee House and the exchange of ideas in early eighteenth century England. In *The Religious Culture of the Huguenots, 1660–1750,* Anne Dunan-Page (ed.). Aldershot: Ashgate Publishing, pp. 163–172.

Hatton, E. (1714). *An Index to Interest.* London: Jonas Brown.

Hatton, E. (1721). *An Intire System of Arithmetic.* London: King.

Hayes, R. (1718). *The Young Merchant's Assistant: or, his Business at the Custom-House Made Easy.* London: R. Hayes.

Hayes, R. (1719). *The Negociator's Magazine: or, the Exchanges Anatomiz'd.* London: R. Hayes

Hayes, R. (1722). *Rules for the Port of London: or, the Water-side Practice.* London: J. Brotherton.

Hayes, R. (1724). *The Negociator's Magazine: or, the Exchanges Anatomiz'd.* London: R. Hayes.

Hayes, R. (1726). *The Money'd Man's Guide: or, the Purchaser's Pocket-companion.* London: W. Meadows.

Hayes, R. (1727). *A New Method for Valuing of Annuities upon Lives.* London: R. Hayes.

Hazen, A. T. (1969). *A Catalogue of Horace Walpole's Library.* New Haven: Yale University Press.

Heilbron, H. L. (1976). Robert Symmer and the two electricities. *Isis* 67: 7–20.

Hérelle, G. (1900). *Documents inédits sur le protestantisme a Vitry-le-François, Épense, Heilitz-le-Maurupt, Nettancourt et Vassy: Depuis la fin des Guerres de Religion jusqu'à la Révolution française.* 3 volumes. Paris: Alphonse Picard.

Herzog, J. J. and Schaff, P. (1908). *The New Schaff-Herzog Encyclopedia of Religious Knowledge*, "Christian Doctrine, Society of," vol. 3, pp. 40–41. S. M. Jackson (ed). New York: Funk and Wagnalls.

Holland, J. D. (1962). The Reverend Thomas Bayes, F.R.S. (1702–1761). *Journal of the Royal Statistical Society (A)* 125: 451–461.

Hoppit, J. (2003). *Parliaments, Nations and Identities in Britain and Ireland, 1660–1850.* Manchester: Manchester University Press.

Howells, C. (1982). *The Kit-Cat Club: A Study of Patronage and Influence in Britain, 1696–1720.* Unpublished PhD thesis, University of California Los Angeles.

Huguenot Society (1914). Conversions et reconnoissances faites à l'Église de la Savoye 1684–1702. *Publications of the Huguenot Society*, vol. 22. London: Spottiswoode.

Huguenot Society (1923). Letters of denization and acts of naturalization for aliens in England & Ireland 1701–1800. *Publications of the Huguenot Society*, vol. 27. Manchester: Sherratt and Hughes.

Huguenot Society (1929). Register of the church of West Street. *Publications of the Huguenot Society*, vol. 32. Frome: Butler and Tanner.

Huguenot Society (1935). Register of the church of Saint Martin Orgars. *Publications of the Huguenot Society*, vol. 37. Frome: Butler and Tanner.

Hunter, M. C. W. (1994). *The Royal Society and its Fellows, 1660–1700: The Morphology of an Early Scientific Institution.* Chalfont St. Giles, England: British Society for the History of Science.

Huygens, C. (1657). De ratiociniis in ludo aleae. In *Exercitationum Mathematicarum libri quinque*, pp. 517–534. F. van Schooten (ed.). Leiden: Elsevier.

Janssens, U. and Schillings, J. (2006). *Lettres de l'Angleterre à Jean Henri Samuel Formey à Berlin de Jean Des Champs, David Durand, Matthieu Maty et d'autres correspondants (1737–1788).* Paris: Honoré Champion.

Janssens-Knorsch, U. (1990). *The Life and "Mémoires secrets" of Jean Des Champs, 1707–1767, Journalist, Minister and Man of Feeling.* London: Huguenot Society.

Jaquelot, I. (1712). *A Specimen of Papal and French Persecution.* London: S. Holt.

Jesseph, D. M. (1993). *Berkeley's Philosophy of Mathematics*. Chicago: University of Chicago Press.

Jordon, C.-E. (1735). *Histoire d'un voyage littéraire fait en 1733 en France, en Angleterre et en Hollande*. The Hague: Adrien Moetjens. Reprinted in 1968 by Slatkine Reprints, Geneva, Switzerland.

Kahle, L. M. (1735). *Elementa Logicae Probabilium Methodo Mathematica in Usum Scientarum et Vitae Adornata*. Halle: Renger.

Keill, J. (1721). *An Introduction to the True Astronomy*. London: Bernard Lintot.

Kippis, A. (1784). *Biographia Britannica: or, the Lives of the Most Eminent Persons Who Have Flourished in Great Britain and Ireland, from the Earliest Ages, to the Present Times*, vol. 3. London: Rivington.

Konnert, M. W. (2006). *Local Politics in the French Wars of Religion: The Towns of Champagne, the Duc de Guise, and the Catholic League, 1560*. Aldershot: Ashgate.

Lalande, J.-J. (1765). Remarques sur la vie & les ouvrages de Mrs de la Caille, Bradley, Mayer & Simpson. *Connoissances des mouvemens célestes,pour l'année commune 1767*, pp. 181–204.

Lalande, J.-J. (1980). *Journal d'un voyage en Angleterre: 1763*. H. Monod-Cassidy (ed.) Oxford: Voltaire Foundation at the Taylor Institute.

Lamy, B. (1701). *Traité de perspective, où sont les fondemens de la peinture*. Paris: Anisson.

Lasocki, D. (1989). The life of Daniel De Moivre. *The Consort* 45, 15–17.

Lasocki, D. (1997). The London publisher John Walsh (1665 or 1666–1736) and the recorder. In *Sine musica nulla vita: Festshrift Hermann Moeck, zum 75*, Nikolaus Delius (ed.) Celle: Moeck Verlag und Musikinstrumentenwerk, pp. 343–374.

Lavedan, P., Hugueney, J., and Henrat, P. (1982). *L'urbanisme à l'époque moderne: XVIe–XVIIIe siècles*. Geneva: Droz.

Le Blanc, J.-B. (1747). *Letters on the English and French Nations*. London: J. Brindley.

Le Gendre, F. (1668). *L'Arithmétique en sa perfection: methodiqment expliquee a la plume et par es ettons, selon l' usage et la pratique tant des financiers que des merchands*. Paris: published by the author.

Le Roy Ladurie, E., Fitou, J.-F. (2001). *Saint-Simon and the Court of Louis XIV*. Arthur Goldhammer (trans.) Chicago: University of Chicago Press.

Lee, W. (1737). *An Essay to Ascertain the Value of Leases and Annuities for Years and Lives, and to Estimate the Chances of the Duration of Lives*. London: S. Birt, D. Browne, and J. Shuckburgh.

Leibniz, G. W. (1716). Methodus Incrementorum Directa & Inversa: auctore Broock Taylor (review). *Acta Eruditorum*, pp. 292–296.

Leibniz, G. W. (1962). *Mathematische Schriften*. Hildesheim: Olms.

Lennon, C. (2006). The Print Trade, 1700–1800. In: *The Irish Book in English, 1550–1800*, pp. 74–88, R. Gillespie and A. Hadfield (eds.). Oxford: Oxford University Press

Lewin, C. G. (2003). *Pensions and Insurance before 1800: A Social History*. East Linton, Scotland: Tuckwell Press.

Lewis, C. T. and Short, C. (1879). *A Latin Dictionary. Founded on Andrews' edition of Freund's Latin dictionary*. Oxford: Clarendon Press.

Lillywhite, B. (1963). *London Coffee Houses*. London: George Allen and Unwin.

Maclaurin, C. (1742). *A Treatise of Fluxions: In Two Books*. Edinburgh: Ruddimans.

Maclaurin, C. (1748). *An Account of Sir Isaac Newton's Philosophical Discoveries, in Four Books*. London: Millar and Nourse.

Maclaurin, C. (1982). *The Collected Letters of Colin McLaurin*. Stella Mills (ed.). Cheshire: Shiva.

Matthews, L. G. (1974). London's immigrant apothecaries, 1600–1800. *Medical History* 18: 262–274.

Maty, M. (1755). Mémoire sur la vie & sur les écrits de Mr. de Moivre. *Journal britannique* 18:1–51.

Mazzone, S. and Roero, C. S. (1997). *Jacob Hermann and the Diffusion of the Leibnizian Calculus in Italy*. Florence: Leo S. Olschki.

McGrath, A. W. (2010). *Science and Religion: A New Introduction*. Chichester: Wiley-Blackwell.

McLachlan, J. O. (1974). *Trade and Peace with Old Spain 1667–1750*. New York: Octagon Books.

Meli, D. B. (1999). Caroline, Leibniz and Clarke. *Journal of the History of Ideas* 60: 469–486.

Messbarger, R. (2002). *The Century of Women: Representation of Women in Eighteenth-century Italian Public Discourse*. Toronto: University of Toronto Press.

Miège, G. (1711). *The Present State of Great-Britain and Ireland*, second edition. London: J. H.

Miège, G. (1711). *The Present State of Great-Britain and Ireland*, third edition. London: J. H.

Montmort, P. R. (1708). *Essay d'analyse sur les jeux de hazard*. Paris: Quillau

Montmort, P. R. (1713). *Essay d'analyse sur les jeux de hazard. Seconde édition.* Paris: Quillau. Reprinted, 1980 by Chelsea Publishing Co., New York.

Monmort, P. R. (1717). De Seriebus Infinitis Tractatus. Pars Prima. *Philosophical Transactions* 30: 633–689.

Morris, G. (1735). *Tables for Renewing and Purchasing Leases*. London: J. Brotherton.

Motteux, P. A. (1740). *Remarques de Pierre Le Motteux sur Rabelais*. London, n.p.

Murdoch, T. (1985). *The Quiet Conquest: The Huguenots 1685 to 1985*. London: Museum of London in association with A. H. Jolly, Ltd.

Murdoch, T. (1992). The Dukes of Montagu as patrons of the Huguenots. *Proceedings of the Huguenot Society* 25: 340–355.

Newman, A. N. (1969). *The Stanhopes of Chevening: A Family Biography*. London: Macmillan.

Newton, I. (1687). *Philosophiae Naturalis Principia Mathmatica*. London: Pepys and Streater.

Newton, I. (1704). *Opticks: or, a Treatise of the Reflections, Inflections and Colours of Light*. London: Smith and Walford.

Newton, I. (1706). *Optice: Sive de Reflexionibus, Refractionibus, Inflexionibus & Coloribus Lucis Libri Tres*. London: Smith and Walford.

Newton, I. (1707). *Arithmetica Universalis; Sive de Compositione et Resolutione Arithmetica Liber*. London: Benjamin Tooke.

Newton, I. (1711). *Analysis per Quantitatum Series, Fluxiones, ac Differentias: cum Enumeratione Linearum Tertii Ordinis*. London: Pearson.

Newton, I. (1718). *Opticks; or, a Treatise of the Reflections, Inflections and Colours of Light*, second edition. London: Innys.

Newton, I. (1722). *Traité d'optique: sur les reflexions, refractions, inflexions, et les couleurs, de la lumiere*. Paris: Montalant.

Newton, I. (1959–1971). *The Correspondence of Isaac Newton*. H. W. Turnbull, J. F. Scott, A. R. Hall, and L. Tilling (eds.). Cambridge: Cambridge University Press.

Newton, I. (1967–1981).*The Mathematical Papers of Isaac Newton*. D. T. Whiteside (ed.). Cambridge: Cambridge University Press.

Newton, I. (1999). *The Principia: Mathematical Principles of Natural Philosophy. A New Translation by I. Bernard Cohen and Anne Whitman*. Berkeley: University of California Press.

Newton, I. and Colson, J. (1736). *The Method of Fluxions and Infinite Series; with its Application to the Geometry of Curve-lines*. London: Woodfall.

Nichols, J. (1812). *Literary Anecdotes of the Eighteenth Century*, vol. II. London: Nichols, Son and Bentley.

Nichols, J. (1858). *Literary Anecdotes of the Eighteenth Century*, vol. VIII. London: J.B. Nichols and Sons

O'Donnell, T. (1936). *History of Life Insurance*. Chicago: American Conservation Co.

Osborne, T. (1742). *A Catalogue of a Choice and Valuable Collection of Books: Being the Libraries of a late eminent Serjeant at Law and of Dr. Edmund Halley*. London, n.p.

Ozanam, J. (1725). *Récréations mathématiques et physiques*. Paris: C. Jombert.

Pearson, D. R. S. (2005). *English Bookbinding Styles 1450–1800: A Handbook*. Cambridge: British Library and Oak Knoll Press.

Peifer, J. (2006). Jacob Bernoulli, maître et rival de son frère Johann. *Journal Electronique d'Histoire des Probabilités et de la Statistique*, 2 (1), http://www.jehps.net/juin2006.html.

Pepusch, J. C. (1746). Of the various genera and species of music among the antients, with some observations concerning their scale. *Philosophical Transactions* 44: 266–274.

Perks, (1706). The construction and properties of a new quadratrix to the hyperbola. *Philosophical Transactions* 25: 2253–2262.

Pitcairne, A. (1693). *Dissertatio de Motu Sanguinis per Vasa Minima*. Leiden: Abraham Elzevier.

Plomer, H. R. (1968). *A Dictionary of the Printers and Booksellers Who Were at Work in England, Scotland and Ireland from 1668 to 1725*. Oxford: TRUEXpress.

Plomer, H. R., Bushnell, G. H., and Dix, E. R. M. (1968). *A Dictionary of the Printers and Booksellers Who Were at Work in England, Scotland and Ireland from 1726 to 1775*. Oxford: TRUEXpress.

Porter, R. (1991). *English Society in the Eighteenth Century*, rev. ed. London: Penguin Books.

Quarrie, P. (2006). The scientific library of the earls of Macclesfield. *Notes and Records of the Royal Society* 22: 5–24.

Record, R. (1662). *Records Arithmetick, or, The Ground of Arts Teaching the Perfect Work and Practice of Arithmetick, both in Whole Numbers and Fractions*. London: James Flesher.

Richards, J. (1730). *The Gentleman's Steward and Tenants of Manors Instructed*. London: John Senex.

Rigaud, S. J. (1965). *Correspondence of Scientific Men of the Seventeenth Century*, vol. II. Hildesheim: Georg Olms.

Roberts, F. (1693). An arithmetical paradox, concerning the chances of lotteries. *Philosophical Transactions* 17: 677–681.

Rocque, J. (1756). *A Plan of the Cities of London and Westminster, and Borough of Southwark*. London, n.p.

Royal Commission on Historical Manuscripts (1910). *Manuscripts of the House of Lords. New Series* 5: 1702–1704. London: H. M. Stationary Office.

Royal Commission on Historical Manuscripts (1912). *Manuscripts of the House of Lords. New Series* 6: 1704–1706. London: H. M. Stationary Office.

Sandifer, C. E. (2007). *The Early Mathematics of Leonhard Euler*. Washington, D.C.: Mathematical Association of America.

Saunderson, N. (1740). *The Elements of Algebra, in Ten Books*. Cambridge: Cambridge University Press.

Schneider, I. (1968). Der Mathematiker Abraham de Moivre. *Archive for History of Exact Sciences* 5: 177–317.

Schneider, I. (2005). Abraham De Moivre, *The Doctrine of Chances* (1718, 1738, 1756). In *Landmark Writings in Western Mathematics, 1640–1940.* Ivor Grattan-Guinness (ed.) .Amsterdam: Elsevier, pp. 105–120.

Scoville, W. C. (1960). *The Persecution of Huguenots and French Economic Development 1680–1720.* Berkeley and Los Angeles: University of California Press.

Senebier, J. (1786). *Histoire littéraire de Genève,* vol. 3. Geneva: Barde, Manget and Company.

Seymour, R. (1719). *The Court Gamester.* London: E. Curll.

Shank, J. B. (2008). *The Newton Wars and the Beginning of the French Enlightenment.* Chicago: University of Chicago Press.

Sherwin, H. (1706). *Mathematical Tables, Contrived after a Most Comprehensive Method.* London: Mount and Page.

Shuttleworth, J. (1709). *A Treatise of Opticks Direct.* London: Midwinter.

Simpson, T. (1740). *The Nature and Laws of Chance.* London: Edward Cave.

Simpson, T. (1742). *The Doctrine of Annuities and Reversions, Deduced from General and Evident Principles.* London: J. Nourse.

Simpson, T. (1743). *An Appendix, Containing Some Remarks on a Late Book on the Same Subject, with Answers to Some Personal and Malignant Representations, in the Preface thereof.* London: J. Nourse.

Smiles, S. (1868). *The Huguenots: Their Settlements, Churches and Industries in England and Ireland.* New York: Harper & Bros.

Smith, D. E. (1922). Among my autographs. *The American Mathematical Monthly* 29: 340–343.

Society for the Diffusion of Useful Knowledge (1837). *The Penny Cyclopaedia of the Society for the Diffusion of Useful Knowledge* 8. London: Charles Knight.

Squire, J. (1742). *A Proposal to Determine Our Longitude.* London: Vaillant.

Statt, D. (1990). The City of London and the controversy over immigration, 1660–1722. *The Historical Journal* 33: 45–61.

Stephen, L. and Lee, S. (1921–22). *Dictionary of National Biography.* London: Oxford University Press.

Stigler, S. M. (1983). Who discovered Bayes's Theorem? *The American Statistician* 37: 290–296.

Stigler, S. M. (1986). *The History of Statistics: The Measurement of Uncertainty before 1900.* Cambridge, MA: Belknap Press.

Stigler, S. M. (1999). *Statistics on the Table: The History of Statistical Concepts and Methods.* Cambridge, MA: Harvard University Press.

Sturdy, D.J. (1995). *Science and Social Status: The Members of the Académie des Science, 1666–1750.* Woodridge, Suffolk: Boydell and Brewer.

Sunderland, C. S. (1881–1883). *Bibliotheca Sunderlandiana: Sale Catalogue of the Truly Important and very Extensive Books Known as the Sunderland or Blenheim Library, sold by Auction by Messrs Puttick and Simpson, London, W.C.* London: Norman & Son.

Tassin, N. (1634). *Les plans et profils de toutes les principales villes et lieux considérables de France.* Paris: Tavernier.

Tardy, L'Abbe (1800). *Manuel du voyageur à Londres; ou recueuil de toutes les instructions nécessaires aux étrangers qui arrivent dans cette capitale.* London: L'Homme.

Taylor, B. (1715). *Methodus Incrementorum Directa & Inversa.* London: Pearson.

Taylor, B. (1719). Apologia D Brook Taylor, J.U.D. & R.S.S. contra V.C. J Bernoullium, Math Prof. Basileæ. *Bibliothèque angloise ou histoire littéraire de la Grande-Bretagne* 6: 403–406.

Taylor, B. (1793). *Contemplatio Philosophica: A Posthumous Work of the Late Brook Taylor LLD FRS*. London: Bulmer.

Terrall, M. (2002). *The Man of Flattened the Earth: Maupertuis and the Sciences in the Enlightenment*. Chicago: University of Chicago Press.

Thatcher, A. R. (1957). Studies in the history of probability and statistics VI. A note on the early solutions of the problem of the duration of play. *Biometrika* 44: 515–518.

Tilmouth, M. (1957). The Royal Academies of 1695. *Music & Letters*, 38, 327–334.

Timperley, C. H. (1839). *A Dictionary of Printers and Printing with the Progress of Literature, Ancient and Modern*. London: H. Johnson.

Tinsley, B. S. (2001). *Pierre Bayle's Reformation: Conscience and Criticism on the Eve of the Enlightenment*. Cranbury, NJ: Associated University Presses.

Todhunter, I. (1865). *A History of the Mathematical Theory of Probability from the Time of Pascal to that of Laplace*. Cambridge: Cambridge University Press. Reprinted 1965, Chelsea Publishing, New York.

Trevigar, L. (1731). *Sectionum Conicarum Elementa Methodo Facilllima Demontrata*. Cambridge, n.p.

Turner, A. J. (1973). Mathematical instruments and the education of gentlemen, *Annals of Science* 30: 51–88.

Tweddle, I. (2003). *James Stirling's Methodus Differentialis: An Annotated Translation of Stirling's Text*. London: Springer.

Tweedie, C. (1922). *James Stirling: A Sketch of his Life and Works along with his Scientific Correspondence*. Oxford: Clarendon Press.

Twiss, R. (1787). *Chess*. London: Robinson.

Venn, J. and Venn, J. A. (1922). *Alumni Cantabrigienses: A Biographical Register of All Known Students, Graduates and Holders of Office at the University of Cambridge from Earliest Times to 1901*. Cambridge: Cambridge University Press.

Vezey, F. (1773). *Cases Argued and Determined in the High Court of Chancery, in the Time of Lord Harwicke, from the Year 1746–7, to 1755*, vol. 2. London: W. Strahan and M. Woodfall.

Vigne, R. and Littleton, C. (2001). *From Strangers to Citizens: The Integration of Immigrant Communities in Britain, Ireland and Colonial America, 1550–1750*. Brighton: Sussex Academic Press.

Voorn, H. (1960). *De Papiermolens in de Provincie Noord-Holland*. Haarlem: Papierwereld.

Walford, C. (1871). *Insurance Cyclopedia*. London: Charles and Edwin Layton.

Wallis, J. (1685). *A Treatise of Algebra*. London: Richard Davis.

Wallis, J. (1699). A letter of Dr Wallis to Dr Sloan, voncerning the quadrature of the parts of the lunula of Hippocrates Chius, performed by Mr John Perks; with the further improvements of the same, by Dr David Gregory, and Mr John Caswell. *Philosophical Transactions* 21: 411–418.

Walpole, H. (1948). *The Yale Edition of Horace Walpole's Correspondence*, vol. 13, W. S. Lewis (ed.). New Haven: Yale University Press.

Ward, J. (1695a). *Synthesis et Analysis. Vulgo Algrebra*. London, n.p..

Ward, J. (1695b). *A Compendium of Algebra*. London: printed for the author.

Ward, J. (1707). *The Young Mathematician's Guide*. London: Midwinter.

Ward, J. (1710). *Clavis Usuræ; or, a Key to Interest, both Simple and Compound*. London: William Taylor.

Waterland, D. (1730). *Advice to a Young Student. With a Method of Study for the Four First Years*. London: John Crownfield.

Weld, C. R. (1848). *A History of the Royal Society with Memoirs of the Presidents*. London: John W. Parker.

Wells, E. (1714). *The Young Gentleman's course of Mathematicks*, vol. 1. London: James Knapton.

Westfall, R. S. (1980). *Never at Rest: A Biography of Isaac Newton*. Cambridge: Cambridge University Press.

White, F. P. and Lidstone, G. J. (1930). A letter of De Moivre and a theorem of Halley. *The Mathematical Gazette* 15: 213–216.

Whitehouse, T. (2010). *Dissenting Education and the Legacy of John Jennings, c.1720–c.1729*. Dr. Williams's Center website: http://www.english.qmul.ac.uk/drwilliams/pubs/jennings%20legacy.html.

Wilkinson, T. T. (1853). Mathematics and mathematicians. The journal of the late Reuben Burrow. *The London, Edinburgh, and Dublin Philosophical Magazine and Journal of Science* 5 (4th series): 185–193, 514–522.

Wingate, E. (1668). *Mr. Wingate's Arithmetick Containing a Plain and Familiar Method for Attaining the Knowledge and Practice of Common Aarithmetick*. Fourth edition. London: Robert Stephens.

Williams, W. P. (1741). *Reports of Cases Argued and Determined in the High Court of Chancery, and of Some Special Cases Adjudged in the Court of King's Bench*, vol. 2. Dublin: Nelson.

Wollenschläger, K. (1933). Der mathematische Briefwechel zwichen Johann I Bernoulli und Abraham de Moivre. *Verhandlungen der Naturforschenden Gesellschaft in Basel* 43: 151–317.

Wood, P. (2003). Science in the Scottish Enlightenment. In *The Cambridge Companion to The Scottish Enlightenment*, pp. 94–116. A. Broadie (ed). Cambridge: Cambridge University Press.

Wordsworth, C. (1877). *Scholae Academicae: Some Account of the Studies at the English Universities in the Eighteenth Century*. Cambridge: Cambridge University Press.

Yonge, W. (1740). *A List of the Colonels, Lieutenant Colonels, Majors, Captains, and Ensigns of His Majesty's Forces on the British Establishment*. London: Thomas Cox, Charles Bathurst, and John Pemberton for the War Office.

Index

Milton Keynes UK
Ingram Content Group UK Ltd.
UKHW020026071024
449327UK00032B/2944

9 780367 382254